IMMEASURABLE WEATHER

ELEMENTS   *A series edited
by Stacy Alaimo and Nicole Starosielski*

IMMEASURABLE WEATHER

METEOROLOGICAL DATA AND SETTLER COLONIALISM FROM 1820 TO HURRICANE SANDY

SARA J. GROSSMAN

DUKE UNIVERSITY PRESS   Durham and London   2023

© 2023 Duke University Press
All rights reserved
Printed in the United States of America on acid-free paper ∞
Project Editor: Michael Trudeau
Designed by Courtney Leigh Richardson
Typeset in Chaparral Pro and Knockout by BW&A Books, Inc.

Library of Congress Cataloging-in-Publication Data
Names: Grossman, Sara J., [date] author.
Title: Immeasurable weather : meteorological data and settler
colonialism from 1820 to Hurricane Sandy / Sara J. Grossman.
Other titles: Elements (Duke University Press)
Description: Durham : Duke University Press, 2023. | Series: Elements
| Includes bibliographical references and index.
Identifiers: LCCN 2023002942 (print)
LCCN 2023002943 (ebook)
ISBN 9781478025023 (paperback)
ISBN 9781478020059 (hardcover)
ISBN 9781478027034 (ebook)
Subjects: LCSH: Meteorology—United States—History.
| Meteorology—United States—Observations—Citizen participa-
tion. | Numerical weather forecasting—Social aspects—United States.
| Weather forecasting—Social aspects—United States. | Climatic
changes—Social aspects. | United States—Climate—Social aspects.
| BISAC: SCIENCE / Earth Sciences / & Climatology
Classification: LCC QC983 .G796 2023 (print) | LCC QC983 (ebook)
| DDC 551.6973—DC23/ENG/20230522
LC record available at https://lccn.loc.gov/2023002942
LC ebook record available at https://lccn.loc.gov/2023002943

Cover art: Star Jet roller-coaster in the Atlantic Ocean from Hurricane
Sandy, Seaside Heights, New Jersey. Courtesy Michael Orso /
Getty Images.

For Grant and Asa

CONTENTS

Water unbuckles the sky. March rain. It feeds the soil, thickens the air, and careens into the rivers, bays, and ocean waters that surround this northeastern region. This is the place—Philadelphia, Lenapehoking—I have called home for most of my life. This homeplace, which has nourished my research and writing, has made my research and writing a home—a place of rejuvenation, invention, safe harbor, and, I hope, generosity. For all that the soil, air, water, and every other unknowable aspect of this environment has given this book and me, I am thankful. And for all that I have taken—space, energy, and more—from this environment to complete my research, I am thankful for such nourishment.

Over the past decade, many people and programs have also propelled this work, making this book richer and fuller with each act of support and care.

Robert Snyder and Rachel Hadas, thank you for being fierce advocates of my work, especially in its early stages. Rob, you helped guide and counsel me through graduate school applications and early attempts at narrative writing and were always eager to share your knowledge of and love for New York City and the Hudson. Thank you for helping me navigate academic culture and for underscoring that environments are precious parts of us. Rachel, you have given me so many gifts that they are hard to count. Your intellectual rigor is unmatched and admirable. Through every poem and essay, you teach how the past leaks into the present, if only we can see it enough to read it. Thank you for shaping my capacity to love this complex world.

Neil Maher, you welcomed me to a new field with excitement and kindness. I take that generosity with me everywhere I go. You have counseled me through this project, chapter by chapter, as well as postgraduate life, with all its ups and downs. Your research and writing guide me, and I am filled with gratitude for all you have given. Thank you for seeing my research and writing for what it was becoming and giving it the space and support to grow.

This book would have been impossible without the generosity of the Smithsonian Institution Archives. Special thanks to Tad Bennicoff, who guided me through the collections during my predoctoral fellowship and many times after! Thanks also to the Smithsonian Institution Predoctoral Fellowship program for funding my research. Kathy Dorman, your precision and care with Joseph Henry's papers were magnificent to witness. I am grateful for the time you took to work with me, including tracing down leads and reading versions of chapter 2. I am indebted to Pamela Henson, whose ever-capacious mind has supported me through years of research. Thank you, Pam, for reading my work, for conferencing with me, for every *Have you read this yet?* . . . , for acting as my adviser, and for opening the doors of the Smithsonian to this poet-historian. This book would have been impossible without you.

Thanks also to the archivists and research specialists at the National Archives, the Library of Congress, and the Smithsonian Air and Space Museum, who supported this work. Special thanks to meteorologist Mark Seiderman at the National Oceanic and Atmospheric Association's National Centers for Environmental Information, Center for Weather and Climate, for providing the invaluable scans that shaped chapter 4.

The Lafayette College Archives, at Sillman Library, was a welcoming place for conducting this research. I am especially grateful for Pamela Murray, who assisted in aspects of research for chapter 2.

Many thanks to the Rutgers-Newark Graduate School and to the American Studies Program at Rutgers-Newark for financial support during my doctoral studies. H. Bruce Franklin, Fran Bartkowski, Jason Cortés, Gabrielle Esperdy, Barbara Foley, James Goodman, and Mark Krasovic were instrumental and supportive advisers of my graduate work. This project would not have grown without the help of fellow graduate students, who not only read and offered comments on many of these chapters but also dedicated themselves to making an intellectual and social community. You are all part of this project in so many ways. Thanks to Erica Tom, Molly Rosner, Katie Singer, Jen Arena, Asha Best, Jules Gill-Peterson, and especially Addie Mahmassani.

Thank you to my colleagues at the Center for Humanities and Information at Pennsylvania State University, who helped to sustain this work and my general well-being from 2016 to 2018: John Russel, Eric Hayot, Pamela VanHaitsma, Anatoly Detwyler, Laura Helton, and Victoria Salinger. Their generous minds helped to feed this book. Thank you also to Bob

Burkholder, Julia Kasdorf, Hester Blum, and Tracy Rutler for their scholarship and warmth at Penn State.

I extend my gratitude to my environmental studies colleagues at Bryn Mawr College—Don Barber, Carol Hager, Helen White, Joshua Moses, and Jonathan Wilson—for giving my research an institutional home and for generously supporting this work in its final years.

I am thankful for Carrie Jackson; Shannon Telenko; and Steve, Fiona, and Will Hathaway. Their generosity through neighborhood dinners and early-morning and late-night chats with or without dogs (though, of course, always better with dogs) sustained me during times of great uncertainty. Gill Street is truly a nourishing place, and I will forever miss the company I had there.

Jennifer Birkeland, you have been an immense support and friend over the years. I would be lost without your insight, humor, intellect, and boldness.

My first animal companion, Sadie, was by my side when I wrote most of this book. My current animal companion, Rex, is by my side as I finish it. Sadie and Rex, you have both taught me more about the way the wind moves and smells than any weather instrument could. The time we have together is never enough.

To the poets, fiction writers, and visual artists who have offered invaluable perspectives for this project, you are here, everywhere. Immense thanks to Armin Tolentino and Michael Johnduff. I am thankful for the friendship and support of Sarah Gerber, Anne Simpson, Bethany Luckenbush, and Erica Ritchie. My dear friends, thank you for all you have offered.

Endless appreciation to Gail, Larry, Alissa, and Dana, as well as Anita, Robyn, and Selma, who have supported me profoundly over the years. Thank you also to Robin, Arthur, and Tappany for their support. Your love is a buoy, and the food we have shared has powered me. Stanley Hochman and Mel Grossman meant the world to me, and the memory of each propels me. Stella Grossman was a loud and proud advocate of my capacity to use my mind—to think and write and find complexity in the world. Even though you are gone, your absolute affection still guides me.

Erin, Dan, Jen, Ben, Denise, and Bob—I am filled with warmth as I think of all the ways you have loved and supported me over the many years we have known one another. From Bob's Party Potatoes™ to all the laughter we have shared, your care has been a bright light in my life.

To Marcella and Anna, your strength and intellects are inspiring. The gift of sisters is a lifelong joy that grows with the years. Josh, Jon, and Matt, I am grateful for each of your tenderhearted natures. Thank you to my parents, Alan and Carol, who have provided for me every step of the way. You were the first who taught me to pay attention to the sky and to the soil—I am forever grateful.

Finally, to my partner, Grant, who has lived this with me. This book grew through dreams, early-morning scribbles, and late-night chats. You have shared, and propelled, all these things and more. Every day now, the world shifts beneath us, and we lean on one another to walk through it—there is snow in April, and there are cherry blooms in December; we eat those late summer peaches with the bruises. You show me all that is good, even though we know that so much is bad. With you, I fall in love with being alive. Finally, finally, to our curious and gentle Asa, my sun, I hope you read these words years from now and know how feverishly you are loved.

The earth has been in an interglacial period since the beginning of the Holocene. The interglacial is a period designation that defines a biosphere by its limited ice sheet cover. As opposed to a glacial period (a time of great ice sheet covering, where significant stores of the biosphere's water are locked in ice), interglaciality defines a globe of mostly, but not entirely, unlocked water, where flows of water move and stray across regions. As the people of the Standing Rock Reservation and their allies have made clear, water is life.[1] The possibilities of the period's watery ways are profound. Many worlds have taken form through the offerings of water, and just as many worlds have been broken or taken away by water's harnessing.

For example, where I live in Philadelphia, on the historical and present-day land of the Lenape, there are rivers and creeks that flood city streets and residents' basements when we get a lot of rain.[2] Waters from Cobbs Creek and Darby Creek join in Southwest Philadelphia, and there they merge into zones of former marshland, the marshland itself abutted by the Schuylkill and Delaware Rivers. When it rains, the water pours. The marshland subearth—on top of which whole communities, and even an airport, rest—bubbles, and the city becomes the sink of the land. The shore seems adrift and, along with it, homes and highways, Superfund sites and sewer waters. But some places get more soaked than others: there, the water is more violent and the weather is uncertain.[3] As climate change makes clear—from Southwest Philadelphia to fifty-eight small, low-lying island nations facing rising waters—the weather is a violent uncertainty. Weather affects all living things, and its impacts are unevenly distributed and exacerbated in regions already facing environmental injustices. Some communities are drowning while others are starved for water.[4]

The geological interval of the interglacial is a scientific designation that underscores the importance of water for life. While the interglacial is a geophysical event with material constitution, it also has, as Kathryn

Yusoff suggests, "social effects." Within the interglacial, we must trace "power relations and relations of force."[5] Certain geosocial formations were chosen, built, and furthered through differing forms of colonialism, from settler colonialism to internal and external colonialisms. Where the Anthropocene is the signature of settler violence in the geological record, the interglacial is the geophysical plane—the stage—on which the Anthropocene has been played. As Heather Davis and Zoe Todd have shown, in its original and now popularized framing, the Anthropocene fails to account for power, differential environmental harms, naming the perpetrators of harm, and holding them accountable.[6] It misses how the geophysical formation of the interglacial has been harnessed to particular world-building ends by settlers, and particularly settler colonialists, as they have used unlocked waters to trespass and lay claim to lands, air, waters, and peoples for settler world-making.

I refer to "settlers" and "settler colonialism" many times throughout this book. When I refer to settlers in this work, I am referring to white populations who historically dispossessed, or who participate in the present-day dispossessions of, Indigenous peoples from their homelands across the United States, and who participated—or whose ancestors participated—in systems of enslavement. In this book, settlers are historic populations (and their present-day descendants) who came to Turtle Island of their own free will, and not under the system of chattel slavery, with the intention to homestead, claim sovereignty over Indigenous lands, and build systems that upheld and perpetuated chattel slavery and Native genocide.[7] Though many settlers see the current flooded, extreme-weathered, and heat-laden world as a new, late twentieth-century occurrence tied directly to the scientific designation of the Anthropocene, the drama of the Anthropocene and its attendant injustices has been playing out for centuries and is, in the US context, directly tied to the sovereignty settlers claim to have over all biospheric life.

Centuries of water move between soil, air, and people over and over again. In a grand converging, the weather mixes everything up; it binds the substrate of the earth and, as this book will argue, binds us to the unexamined and harmful settler worldviews that have shaped environmental data collection in the United States. Weather knowledge is a kind of knowledge that can both sustain and shatter life. This book focuses on the weather knowledge that was produced through data gathering, and practices of quantification more generally, in the US settler states from roughly 1820 to the time of Hurricane Sandy in 2012 and reveals that weather cannot be assumed as a universal experience, set of phenomena, or collection of captured data. *Immeasurable Weather* takes up weather-dependent settler colonialism and rescripts settler environmental data collection as part of the larger project of settler colonialism in the US settler states. I argue that data gathering and quantification gave coherence to a national weather project, and to a notion—grounded on settler and heteropatriarchal values—of "nation" itself. Further, this book elaborates settler complicity in assembling and upholding an instrumentalized view of weather with the goal of settler world-building at all costs. This included dominance and dispossession of Indigenous lands and waters, genocide, and the building and perpetuation of systems of enslavement and white dominance. Much as Traci Brynne Voyles has done in *The Settler Sea* with regard to the Salton Sea, this book "examines the ways that settlers maintain, shape, manage, and mismanage the nonhuman world . . . restructur[ing] physical landscapes in ways that exert and reinforce processes that are part and parcel of colonial power relations in the United States."[8]

Settlers utilized notions of data to disentangle themselves from networks of relationality, responsibility, and accountability. This separation, explored throughout the book, is a loss that can be felt on multiple levels. It can be felt in terms of the effects that petrocapitalism has had on frontline and high-risk communities and more-than-human members of the biotic world. This separation resounds in the ways that settlers themselves are bereft of their own relationality in historical memory, practicing a collectivized forgetting of the violence they perpetrate, and have perpetuated, in the name of settlement, whiteness, and capital accumulation. This book studies the emergence and evolution of weather data from and within the frame of local, settler data making in a US context across the nineteenth and twentieth centuries. In refusing to study settler science from a "universal position" and in seeking to "write from where I know," I uncover the individuals, communities, and ideologies that have built and informed settler weather science in the United States from the early nineteenth century to the present.[9] This is not a history of institutions, although institutions are central characters. It is a history of networked settler volunteer communities and their popularization of data as a fundamental category within the settler environmental imagination.

The following examples—which are in no way exhaustive—reveal how the material condition of a geophysical world of unlocked water was harnessed to enact harm while traveling hand in hand with scientific, and often meteorological, capture: (1) The British Empire utilized water and weather data for territorial expansion when, throughout the eighteenth century, the East India Company kept logbooks of the weather from 1786 to 1834 on "empire voyages" from southern English ports and the east coast of India, Indonesia, and southeast China.[10] (2) The Dutch utilized weather journals on their colonization crusades to Java and elsewhere.[11] (3) The United States used military forts across the continent as weather data collection stations, which aided in wartime and nation-building terrain advancement, Indigenous dispossession, and settler world-building.[12] (4) The US Air Force, and specifically the Air Force Fourteenth Weather Squadron, collected upper-air radiosonde data from Iraqi Meteorological Department launches across the period of 1958–89, which were used "as reference for flights during Operation Desert Storm in 1990–1993."[13]

The history of science's relation to injustices such as settler colonialism and capitalism in the US settler states has been explored through the work of scholars Tony Perry, Emily Pawley, Paul Lucier, and Conevery Bolton Valenčius. In his article "In Bondage When Cold Was King: The Frigid Terrain of Slavery in Antebellum Maryland," Perry elaborates on "the ways enslaved people leveraged cold weather against slaveholders," revealing how enslaved people and slaveholders "differently mobilized wintry environments against one another in contests over power." In detailing the ways that "slaveholders and the enslaved leveraged elements of the dynamic environment against one another in efforts to assert their respective wills," Perry's research attends to the weathery power relations within the Little Ice Age of antebellum Maryland.[14] In *The Nature of the Future: Agriculture, Science, and Capitalism in the Antebellum North*, Pawley documents the beliefs and practices of "improving agriculturists" in the antebellum North, who approached nature as a "balance sheet." In studying the agriculture of New York, Pawley shows how settlers made nature "legible, measurable, and above all calculable" for the purposes of capital gain.[15] Lucier has documented the ways that scientists have acted not as stewards of the biosphere but as extensions of the capital marketplace. In *Scientists and Swindlers: Consulting on Coal and Oil in America, 1820–1890*, he explores the relation between American earth scientists and the mining industries of the nineteenth century.[16] Valenčius argues that nineteenth-century settlers experienced a "profound connection" with the western agricultural landscapes they sought to inhabit and that settlers' own bodies were sites of environmental and medical connection with the lands they occupied.[17] Building on this work, *Immeasurable Weather* attends to the interplay of settler colonialism, patriarchal power, and whiteness across a two-century history of settler investment in meteorological data culture.

Additionally, while the relation between Enlightenment science and nation-building has been well attended to in the British, German, and Australian colonial periods, this kind of work has not been done for the post-1800 United States.[18] Jan Golinski has revealed that the origins of weather observation in Enlightenment Britain date back to 1640. Although weather observation was not an official branch of the nation-state, it was part of public culture: weather observation was an important practice for conquest, as knowledge of climate conditions in occupied territories shapes colonial power.[19] In the US settler states—the focus of this book—weather observation functioned similarly as a form of deterritorialization; it also connected settlers to the idea (and the locally situated practices) of a scientific nation-state.

Settler colonialism posits a particular relation to land and is, as Patrick Wolfe argues, a "land-centred project."[20] As Eve Tuck and Marcia McKenzie note, "Settler societies are based on ongoing displacement and dispossession of people in relation to land" and are founded on "deep behaviors of ignorance toward land, water, environment, and sustainability, as evidenced in fuel extractions, agricultural practices, pollution and toxic dumping, hyper-development, and water use."[21] For settlers, weather data enacted settler worldviews that environments—from weather to water to biota—were objects to be measured, with their meaning and value resting in their capacities to be known through quantification. Weather data obfuscated relationality. Relationality—or the lack thereof—is a major theme in this work. By relationality, I mean to analyze the ways that settlers historically related to the more-than-human environments they occupied—from soil to water to air—across the US settler states.[22]

Importantly, settlers did not think of land as *Land*—as Sandra Styres, Celia Haig-Brown, and Melissa Blimkie describe it—an "animate and spiritual being constantly in flux," which "encompasses all water, earth, and air."[23] Following Styres, Haig-Brown, and Blimkie, as well as Max Liboiron, the capitalization of the word *Land* here refers to "a proper name indicating a primary relationship," whereas "small-*l* land" refers to "the concept from a colonial worldview, whereby landscapes are common, universal, and everywhere, even with great variation."[24] Throughout this book and the history it tells, I use *land* in the nonproper sense to indicate the relationship settlers had with the Lands they occupied. Liboiron continues, "Calling out proper nouns so they are also proper names is part of a tradition where using someone/thing's name is to bring it out of the shadows and engage it—in power, in challenge, in recognition, in kinship."[25] While I recognize the Land that has allowed for my research to take place—that has nourished and sustained me, as well as comforted me in times of uncertainty—I also recognize that the historical actors of this book act toward land and not Land. That is an important distinction that has an impact on how these historical actors were able to conceive of relationality.

While nineteenth-century settlers understood there was a deep connection between the health of their bodies and the lands they occupied, settlers' own understandings of body, land, and health were deeply entrenched in ideals such as "freedom," as well as their belief that the land was theirs to settle and exploit through agriculture.[26] Their way of relating to the land—even when through the "health" of their bodies—was uniquely tied to the logics of settler colonialism, which I argue separated "knowledge from responsibility," in the words of Robin Wall Kimmerer.[27] Part of the journey of this book is to uncover how data culture separated the knowledge of weather and climate from responsibility. Across the book, I refer to the US settler states as a nation-state. When I use this language, I am referring both to material processes (such as the building of settler infrastructures like railroads, forts, and data centers) that allowed for the proliferation of settlement and dispossession and to the ideas (such as domination of environments and property ownership) about the land and the weather that shaped these processes.

Climate change is the logical product of the ways that settlers have related to the environment through data infrastructures.[28] Settler data culture was built by separating knowledge from responsibility, and this relation helped conceive of, project, and further American weather culture. Data culture stabilized a form of relating to environments that isolated the weather—and other parts of the world—as objects to be studied rather than as subjects to be respected and understood through systems of reciprocity and care. This made several ways of life possible for settlers. First, by upholding quantification as the primary means for knowing and making environmental knowledge in scientific spheres, settlers practiced erasure. Their prioritization of Enlightenment measurement schemas marginalized Indigenous environmental knowledges dating back to time immemorial. Traditional ecological knowledge (TEK) holds various understandings of weather's relation to living systems, human and nonhuman. Raymond Pierotti and Daniel Wildcat write that TEK might be considered "an intellectual foundation for an indigenous theory and practice of politics and ethics, centered on natural places and connection to the natural world."[29] Importantly, as Heather A. Smith and Karyn Sharp point out, TEK is a contested concept, which has "many definitions" and for which "there is no consensus of an operational definition applicable across disciplines."[30] These knowledges existed long before settlers began measuring the weather in this period, and they continue to exist well beyond nineteenth- and twentieth-century settler colonial expansion.[31] Second, as they continued to build data culture, settlers engaged in a collective forgetting that "obscure[d] and eviscerate[d] memory" of colonial violence—genocide, forced removal, and settler colonial "education" systems—that made scientific measurement possible. This "foreclosure of knowledge" produced a sense of history with "all harm, trauma, and associated accountability" for settler colonial violence firmly rooted in the past. This resulted in a settler subjectivity that was and remains occupied with "the settler as innocent and desiring of democracy, fullness, and opportunity."[32] These foreclosures, and the subjectivities they created, are not separate from the history and culture of scientific knowledge making in the US settler states, but rather are origin points for the evolution of meteorological science itself.

This book treats the ubiquity of settler practices, technologies, and their links to a pervasive understanding of white supremacy as the backbone for meteorological data production in the US settler states. From these things grew not only meteorological science but also the notion of a nation grounded in a subjectivity of whiteness. As Audra Mitchell and Aadita Chaudhury elaborate, whiteness can be understood as "a set of cultural, political, economic, normative, and subjective structures derived from Eurocentric societies and propagated through global formations such as colonization and capitalism."[33] Understood in this way, whiteness is created through "multi-scalar structures," is not "reducible to skin pigmentation, genetics or genealogy," and "is remarkable in its ability to render itself invisible to those who possess and benefit from it."[34] Scientific practices, the building of settler subjectivities, and the production of the project of "American" weather were integral to the larger purposes of settler colonialism and whiteness.

While this work focuses on settler colonialism as a structure that is historically "land-centered," Tiffany Lethabo King has elaborated on the ways that the preoccupation within settler colonial studies "with the settler's relationship to land, or terra nullius . . . displaces how settlers also become conquistadors/(humans) through Native genocide and Black dehumanization."[35] Rather, King argues that "conquest," as opposed to the settler's relationship to land, "is a larger conceptual and material terrain than settler colonialism and far more suited for the regional/ hemispheric particularities of coloniality in the Americas."[36] Following King, this book does try to attend to the ways conquest and land theft were interlinked processes that aided the development of meteorological data collection and the development of a national meteorological science. However, this book's dominant focus on white settler scientific production, or "the elucidation of the settler and their concerns," risks "interrupt[ing] [the] examination of the violence of the slave trader and serial murderer of Indigenous people."[37] This underlying tension is part of the book and has not yet been resolved. In marking the nexus points of white settler scientific world-building, my intention is to allow for other ways of conceiving of and relating to the biosphere that exceed settler colonial and conquest-based structures.

As settlers learn to float in a world we have sunk, we must reckon with the ways that, to return to Kathryn Yusoff, "different forms of inequality have result[ed] from the differential harnessing of these geopowers."[38] The environments we inhabit and the measurement systems we have built and continue to build have been used by us, and our communities, in a type of settler world-building, which has structurally disenfranchised Black and Indigenous communities, as well as many others. The Anthropocene does not signify humans' altering of the geological strata, but is rather the material articulation (the living sediment) of what settlers and "petrochemical companies and those invested in and profiting from petrocapitalism and colonialism" have enacted in and across the interglacial.[39] Data capture was, and continues to be, a central logic in this interglacial world, a logic that fuels the writing of the Anthropocene strata. The history of data's emergence and messy evolution in the melting zone of the interglacial period was and is uniquely bound to settlement ideology, an ideology that is literally deadly—an ideology that counts the biosphere, and the life within it, as capital. In the United States specifically, the "language of numbers" was an "enumerative strategy" that buttressed and created a coherent settler colonial space.[40] *Immeasurable Weather* asks what interglacial data making looks like in the settler context and traces the differential knowledges that data making has proliferated in this context.

IN HIS 1856 SECRETARIAL REPORT for the Smithsonian Institution, Joseph Henry pleaded for public patience. He was overworked and overwhelmed, and he needed help processing the stacks of meteorological data sheets that had been accumulating for almost a decade at the Smithsonian's Meteorological Project office in Washington, DC. Henry's team had been steadily reducing incoming data since 1849—all sent from Smithsonian Meteorological Project volunteers across the United States—but by 1856 there was simply too much data for Henry and his team to process, reduce, and send to print. This was a problem not only for the Smithsonian but also for the public it served, which had developed an insatiable appetite for weather data over the last half century. Along with a trusted friend and scientist, James Coffin, a professor of mathematics at Lafayette College, Henry came up with a plan. These two white men decided to hire white women—whom, they agreed, they could pay significantly less than men—to solve the problem of data overload. By 1857, "twelve to fifteen persons, many of them females," were employed as human computers to reduce meteorological observations by converting numerical weather observations from across the country into data sets.[41]

This nineteenth-century intersection of settler science, an overwhelmed institution, and gendered labor sits at the origin point of contemporary US understandings of weather, data, and climate. This book charts the two-century history of weather data by investigating the labors associated with bringing data to life, the diverse media forms that data took, and the social embeddedness of environmental science in settler colonial communities. Roughly two hundred years before "big data," mid-nineteenth-century human computers and meteorological volunteers interfaced with locally situated weather reports, which would gradually become significant and impactful climate data.

In the mid-nineteenth century, data collection shifted from an East Coast practice to one that attempted to encompass new settler territory as far as the ancestral lands of the Ramaytush Ohlone, or what settlers call San Francisco.[42] The first settler weather observations from this territory were taken by Lieutenant John C. Frémont of the US Army in 1845, five years before California's recognition a state. First as invading armies, and then as "volunteers" and "experimenters," different formations of settlers from across the nineteenth and twentieth centuries aided the US government in creating a vast, relational weather database in which all historic and contemporary data points could be put into conversation. Over two centuries, settler experimenters, government employees, volunteers, and institutions used the language of data to produce and share "knowledge of our climate in all its relations," as Joseph Henry described in 1852.[43] They used data to create a coherent national space that foretold settler futures and settler pasts through data archives.

This book centers around key moments of environmental data configuration, rupture, and crisis across two centuries. I tell an intergenerational story of data labor, with specific attention to how these labors nurtured ways of understanding one of the most important forces shaping everyday life: the weather. My research locates the rise of data as a concept and practice and then documents its monumental shifts across the technological landscape of the last two centuries. Tables and graphs were tools of settler colonial territorial advancement and world-building. Throughout the book, I study how correspondence, scientific journals, instruction manuals, reports, manuscripts, and newspapers from the Smithsonian Institution, the National Archives, the National Aeronautics and Space Administration, and National Oceanic and Atmospheric Association (NOAA) research centers articulated a national concept of data that allowed for the imagination, sustenance, and proliferation of settler worlds. My broad analysis of these archival documents provides the backdrop for detailed portraits of five different communities of weather data laborers.

As a settler scholar living and working on the Lenape lands, I struggle with the stories of weather science and environmental data capture that have been told by settler scholars. These stories that settlers tell themselves about weather science are rooted in notions of discovery, experimentation, and seemingly democratic or grassroots science.[44] But they pay no mind to the ways that "grassroots," or "citizen," science was enacted in the early to mid-nineteenth-century period by non-Indigenous peoples in order to claim "settler sovereignty over all things"—specifically, land, air, water, and the "subterranean earth."[45] These stories of Enlightenment objectivity and scientific progress have been incredibly mobile within the history of science, within environmental history, as well as in science and technology studies. If we tell ourselves stories in order to live, as Joan Didion once said, then we need to interrogate the kinds of living that these stories in particular make possible. My work investigates how, in the United States, quantifying the weather was tied up in existing injustices such as settler colonialism, patriarchal power, and white supremacy.

Weather science and data collection emerged as a practice and flourished in the nineteenth century while several communities faced injustice and violence, some through forced removal from their ancestral lands and others through forced enslavement. "Citizen science" traveled hand in hand with settler territorial aims and claims, patriarchal power, and white supremacy.[46] Whiteness was a powerful category across the nineteenth century, as Gregory Smithers argues,[47] and well before then, as Cheryl Harris has shown. Harris points to the ways that the social construction of race was active as early as the seventeenth century, where "laws parceled out differential treatment based on racial categories." Smithers argues that the category of whiteness was contested across the nineteenth century.[48] Yet, I argue, despite this categorical contestation, whiteness shaped power structures in this period and much earlier; as an orientation, Sara Ahmed writes, whiteness "puts certain things within reach . . . not just physical objects, but also styles, capacities, aspirations, techniques, habits."[49] My research takes seriously the phenomenological and ideological power of whiteness across this period, noting that while some settlers articulated themselves through discourses of "whiteness," documents from settlers more often make "implicit race and gender dimensions."[50] Understanding oneself as white was not, in fact, the primary way in which whiteness acted historically. Ahmed further suggests the operation and coherence of whiteness as an "unnoticing." Unnoticed whiteness by white people is part of its power—the fact that white bodies do not have to face their whiteness is the operation of its privilege.[51]

Reporting weather data to institutions throughout the nineteenth century required an address. This means that performing science was tied up in land occupation and the rights of property ownership, which was afforded exclusively to white settlers at the time.[52] The National Centers for Environmental Information, part of NOAA, categorizes nineteenth-century weather data from the unceded lands of the Dakota and Lakota peoples as "climate data records" without any acknowledgment that these data were gathered by generals who were occupying such lands.[53] Settlers utilized these occupations not only to overharvest habitats and animals, which were crucial to Indigenous—here, Dakota and Lakota—life but also to gather meteorological data to send back to the federal government, data that furthered settler knowledge of local climate and terrain, and thus advanced settler movement and the creation of settler infrastructure in the region. The violence of settler appetites—and especially settler institution appetites—for environmental data has not faded over time but rather has intensified, as the recent Mauna Kea Hui resistance to the Hawai'i State Land Board and the University of Hawai'i's plan for an astronomical telescope on the sacred lands of the Mauna Kea shows.[54] While scholars of science, technology, and society have been studying the sociocultural dimensions of data big and small by attending to the history of cybernetics and nineteenth- and twentieth-century human computing, such studies rarely explore the environmental dimensions of data, or data as a form of settler colonial world-building. Additionally, though historians of technology have begun to ask questions about twentieth- and twenty-first-century labor, from YouTube content moderation to "gendered technocracy" in British computing, these inquiries into data labor have yet to tackle environmental data in particular.[55] At the same time, critical inquiries into the history of information have not played a major role in the scholarship of settlement within the field of environmental history to date, nor has the relationship between settler colonial world-building and Indigenous persistence and erasure been fully addressed.[56] Historians of climate change have begun to historicize information graphics, but their lack of attention to the cultural production of data itself has left us with a dim understanding of cultures of environmental data as well as that culture's emergence among a gradually professionalizing scientific community that included underpaid women in the nineteenth century.[57] *Immeasurable Weather* bridges the gaps between these fields by centering an environmental humanities lens that attends to the hinge space between these fields.

Immeasurable Weather is a necessary book for today's world not only because it bridges these disciplinary gaps but also because it illuminates why all the data in the world will not be an elixir for extreme weather, environmental collapse, and climate change. Though settler scientists and environmentalists imagined that more environmental data would increase environmental action and change, quite the opposite has happened. *Immeasurable Weather* explains this phenomenon through a sociocultural history of environmental data, situating the close reading of primary source materials—from letters enclosing live specimens like locusts to hand-filled data forms—alongside cornerstone questions belonging to environmental and information studies. I track how weather has become a numerical experience, a form of data itself, and I ask how US communities in the nineteenth and twentieth centuries perceived the weather as a distinctly numerical phenomenon, why it was so important to do so, and the inheritances of that thinking. In asking these questions, I assert the link between settler colonialism and top-down as well as bottom-up scientific practice. I frame settler environmental data as resonant for media and information studies, history of science, and especially environmental media studies.

In the US settler states, data emerges and gains popularity in the specific context and evolution of American weather. In exploring the rich environmental origins of data, each chapter illuminates the power that everyday communities have in shaping data. First, I argue that between 1820 and 1890, data was conceptualized by a public sphere whose efforts would lay the groundwork for the National Weather Bureau and, later, the National Weather Service. Settler farmers and teachers were building conceptual frameworks for American data in this period—frameworks that were centered on tactility—but they were also the original advocates for and compilers of American climate archives. Next, I argue that, from 1890 to 1940, data was reconceptualized once again when Weather Bureau workers and Dust Bowl weather volunteers are tasked with data stewardship in a landscape of automated instrumentation and weather crisis. Finally, with the rise of meteorological satellites in the 1960s, data is more formally enmeshed in military-meteorological nation-state structures. As such, data moves away from locally situated systems of caretaking and toward instrument-oriented and semigalactic perspectives that further obscure settlers' bonds to the environments they inhabit. This obscuration is often crystallized during crisis events, such as Hurricane Sandy, the final focus of the book's concluding chapter. In exploring these formative periods, I highlight the ways that humans have been producers, computers, compilers, and caretakers of data. I explore what these relationships with data have meant historically, how they can be used to contextualize data today, and how they might be employed to transition to a more just environmental future.

The history of data, the pervasiveness of data, and data's relation to injustice have been attended to by numerous scholars, including Catherine D'Ignazio and Lauren Klein, Lisa Gitelman, Rob Kitchin, Kathleen Pine and Max Liboiron, David Ribes and Steven Jackson, and Daniel Rosenberg.[58] D'Ignazio's and Klein's *Data Feminism* positions data as "a double-edged sword," where "data is power" and "the power of data is wielded unjustly."[59] Gitelman understands data as a cultural form, made and remade through communities of practice.[60] Kitchin treats data as a recent historical manifestation, offering definitions of big data as well as advice for harnessing data in contemporary research landscapes.[61] Pine and Liboiron attend to the "measurements, the artifacts and practices that form some of the smallest units of quantification underlying data, algorithms, and other vectors of representation" that "power dynamics, knowledge systems, and culturally-based assumptions."[62] Taken together, Gitelman, Kitchen, and Pine and Liboiron help question notions of raw data, positioning data as a product of social and cultural factors. Rosenberg adds the understanding that "data" itself was shifting in meaning across the nineteenth century, operating at the beginning of the century as the "premise" of investigation but as the "result of investigation" at the end.[63]

Meanwhile, in histories of science, scholars have rightfully addressed data gathering and environmental display as sociocultural processes. Lorraine Daston and James R. Fleming show how scientific observation—a form of data gathering—emerges across institutional networks.[64] Much like Gitelman and Kitchen, and Pine and Liboiron, historians of science also illuminate the situated practices that render data a product of culture. While environmental history scholars have been less concerned with data proper, Kristine Harper and Joshua Howe have contributed to our understandings of environmental data display.[65] Harper considers the art of meteorology within the history of meteorological computation, arguing that computational display was both a science and an art in the postwar period. Similarly, Howe explores theories of the natural lurking behind environmental displays like the Keeling curve (a graphical representation of atmospheric CO_2). Taken together, these works develop a rich, transdisciplinary approach to the study of data concepts and observational and display practices. My book enriches this approach by offering a deep history of weather data culture, asking how data itself grew roots across the US settler state and how these roots have spread over the last two centuries.

As this book demonstrates, weather data recording and measurement have deeply shaped settler environmental consciousnesses, access to land and Land, and the capacity for settlers to narrate their past and imagine a future.[66] In this book, I isolate the ways that settlers in the United States have come to understand the weather through systems of numerical capture (today is seventy-eight degrees with fifteen-mile-per-hour winds from the west), and specifically the ways that these numerical systems mobilized settler perspectives and worldviews. Settlers crafted a national concept of data, and this concept organized settler worlds around the quantification of environments, leaving little room for relations of reciprocity and responsibility with the worlds they inhabited.

CHAPTER 1 BEGINS with the story of two volunteer observers in Ithaca, New York, and their daily routine of observing the weather in the mid-1820s. The chapter then expands out to consider a regional network of newly titled "weather observers" across the northeastern United States and the ways in which this network utilized weather measurement to uphold settler claims to territory and build archives of settler environmental knowledge. Data collection in this period grew out of land surveying, which furthered Indigenous land theft: collecting "national" data meant extending and occupying unceded territories. The network of observers who furthered settler surveying and data collection included members of institutions such the New York State Regents, the Franklin Institute, Girard College, and the Maryland Academy of Science and Literature. I show that as settler volunteers around the eastern United States collected daily weather observations, they created an origin story for settler environmental knowledge. Over the course of the early nineteenth century, these observations would become a collection of regional meteorological data points. In doing this, they depended on the graphic form of the table for storing and processing data. By studying this work, I chart the rise of data practices and data language within settler meteorological communities of the early century.

With these practices and discourses in place, the stage was set for a meteorological observation project that would seek to encompass the transitional national boundaries. As plans for a national project brewed through the efforts of two seminal figures—Joseph Henry and James Espy—observers and institution officials positioned and solidified data as a public good: observers and institutions managed the practice of weather observation, assigned the language of data to collections of observations, and, finally, joined data discourses with calls for public data and climate archives. I argue that this process illustrates how data was conceptualized through the social life of settler science. These practices worked to inform settler environmental imaginations, positioning weather and environments as knowable through the processes of quantification. Much like landownership, data traveled with notions of settler patriarchal dominance and whiteness.

Whereas the work of male weather enthusiasts—and the tables they created—helped to discipline data in the early part of the century, chapter 2 tells an allied story of data's move from regional importance to national importance in the midcentury period through the labors of white women. This chapter uncovers the work of several groups of white women who acted as weather data collectors and meteorological calculators for the Smithsonian Meteorological Project, the first settler-operated national weather data collection project in the United States. Working with correspondence and institutional publications, I show how women volunteers understood themselves as weather data laborers, how they utilized patriarchal power to navigate male meteorological data culture, and the ways they were met with resistance and erasure under that same patriarchal system.

This chapter travels from Washington, DC, to Easton, Pennsylvania, to Nebraska Territory and back again. The historical characters include famous women, like Clara Barton, as well as women computers whose work shaped scientific practice in the mid-nineteenth century but whose names could not be recovered. The named and nameless women volunteers, computers, and copyists who were part of this project helped to build the foundations of a settler colonial meteorological data collection and computation system in the United States. I argue that though the Smithsonian Meteorological Project was a unique opening for women to enter scientific roles, weather data collection and calculation practices in the mid-nineteenth century were tied up in cultural power structures that made it impossible for women to practice weather data collection and calculation as men could and would at the time. These power structures positioned masculine skill at the center of meteorological data labor and women's data labor at its edge. The chapter shows that white women were able to enter meteorological science only by furthering the intertwined projects of settler colonial expansion and white male supremacy, a supremacy that sought simultaneously to utilize and erase their efforts.

Chapter 3 considers how the utilization of meteorological kite technology facilitated the search for upper-air data from 1890 on. In this chapter, I explore the ways that data was reconfigured through emerging automatic and remote technologies that combined with meteorological kite experimentation. Beginning with weather experimenter W. A. Eddy launching a kite, with a camera mounted to its frame, into the air on the shores of Bayonne, New Jersey, the chapter captures the experimenter-based nature of upper-air exploration at the turn of the century. Weather kites not only allowed for the collection of upper-air data but also acted as a popular toy-turned-tool that propelled white male communities into professional science. This chapter follows data collection as it moves from the ground to the air through amateur weather-kite tinkering along the Eastern Seaboard from 1880 to 1900, examining how aerial data exploration depended on the cultures of masculine science and settler masculinities. I outline the ways that white masculinity extended US military dominance through formal and informal partnerships with male weather experimenters.

The popularity of weather kites mounted with cameras, thermometers, and barometers literally shifted perspectives, physically reorienting where experimenters—and later institutions—gathered meteorological data as well as the ways that consumers of data understood the weather around and above them. By the late century, remote sensing technologies combined with pictorial imaging, as exemplified by emerging aerial photography, drawings and lithographs published in newly popular scientific journals. Photographs were understood as scientific and quantitative material that visualized the urban landscape with great clarity. Print culture supported the formation of a settler-defined meteorological knowledge that was not only popular but also textual and graphics-based, white, and male. These new media produced a national weather consciousness (more on that later in this introduction) rooted in aerial technology and the promise of endless quantification of aerial space through the dominance of pictorial data. This period begins to lay the groundwork for satellite meteorology in the 1960s.

Chapter 4 opens with a description of Weather Bureau employee George Franklin and his office at 102½ Spring Street in Los Angeles, California, on the morning of June 6, 1900. As a Weather Bureau employee, Franklin was responsible for an ensemble of self-recording weather instruments. His work represents an important shift in data labor as weather data collection becomes systematized and automated throughout the early part of the twentieth century. Franklin and others acted as caretakers of data rather than its primary producers, and across the nation, publics became consumers of weather data products that relied on pictorial interpretations of weather data rather than its numerical presentation. This chapter shows how self-registering instruments altered data relationships by restructuring notions of data as anonymous, autonomous, and pictorial for producing and consuming publics. Through a study of instruction manuals and Weather Bureau reports, I explore how the bureau and its personnel created data qualification schemes that positioned data between two poles—"clean" and "good"; "dirty" and "bad."

As the Weather Bureau continued to craft data collection practices, professional meteorology was also on the rise. Chapter 4 focuses on the Weather Bureau within the context of the rise of professional meteorology and pinpoints how discourses of clean/good and dirty/bad shaped data classification systems and environmental consciousness in general. The chapter also explores how, between 1900 and 1930, data landscapes grew more complex as self-registering instruments produced twenty-four-hour data streams, and bureau employees, volunteers, and publics grew estranged from the data they were once charged with creating a century earlier. This estrangement is highlighted in the chapter's consideration of the Dust Bowl. An examination of this event shows how data measurement and management regimes failed to render the escalating catastrophe of airborne dust and soil erosion, how dust data fell out of existing data-capturing systems. The onset of dust challenged existing modes of data collection and expression. Importantly, it challenged the efficacy of settler data measurement systems and the promise of endless quantification. In the wake of the Dust Bowl, World War I, and World War II, endless and constant data streams were realized through satellite meteorology. Twenty-four-hour data streams—from ground-based sensors to satellite feeds—could narrate environmental pasts for the nation and foretold a quantifiable future. There was more to know—and leverage—about American weather, and meteorological satellites were essential to that knowledge.

The role of satellite meteorology, and the constant access to land (as resource) and life (as data point) such technology assumes, is the focus of *Immeasurable Weather's* final chapter.[67] In the two-century drama of weather data's emergence and utilization in the US settler states, satellite data becomes a restructuring force in the final act. I begin chapter 5 by documenting the emergence of satellite meteorology through cooperative efforts by the War Department, the Weather Bureau, the Central Intelligence Agency, and the RAND Corporation. From TIROS to NIMBUS, and Landsat to GOES, satellite meteorology was a product of collaboration between these sectors and the various desires (military reconnaissance, resource extraction, and national science) they held. Satellite meteorology consolidated the nation-state's power across the middle to late twentieth century, yet it has also left inhabitants of that same nation-state bereft. Surrounded by data, they could not see themselves amid it.

The second section of chapter 5 opens on October 26, 2012, three days before Hurricane Sandy's East Coast landfall, when NASA released its first aerial, time-lapse video of the storm. Roughly twenty-two thousand miles above earth, NOAA's GOES 14 satellite scanned the hurricane's movement across the southern United States. The NASA video, produced through rapid-scanning technology, captured the enormity of the storm, its dynamic cloud structure, and the sharpening of the storm's eye as the continent dimmed under the setting sun. But despite the complicated process of data gathering and transmission that is involved in meteorological satellite sensing and imaging, this satellite media passed as uninterpreted weather reality for popular audiences, offering visual and textual narratives of weather crises that abstracted viewing publics from environmental disaster. Contemporary consumers of weather data cannot see the back-end processes that form popular weather and climate representations. By analyzing the meteorological satellite movement of the 1960s, contemporary GOES 14 experimental satellite technology, and visual satellite data and media during storm crises, I argue that weather data visualizations appear as uninterpreted environmental realities and that this has magnified an already existing tear between weather systems and species that live and die by weather's logics.

Immeasurable Weather argues that data is made and unmade in the local and the everyday; it is a tool that is as dynamic as it is contested. Data is steeped in traditions and structures that have profoundly impacted the stories settlers tell about the environments they occupy. "Stories are compasses and architecture," writes Rebecca Solnit. "We navigate by them, we build."[68] It is time, and has long been, to reckon with landscapes built from stories of environmental measurement—to reckon with the architecture, and to ask what a turn to transformative environmental justice could allow considering this history.

This book charts how, across two centuries, weather data was an outgrowth of settler colonial ideologies of domination and atomism. These ideologies not only informed environmental measurement but also built a settler sense of place that understood biospheric systems as categorically distinct rather than as nodes within a larger interconnected and variable earth network.[69] These systems of data capture informed how settler data collectors understood the crisis itself. A brief visual his-

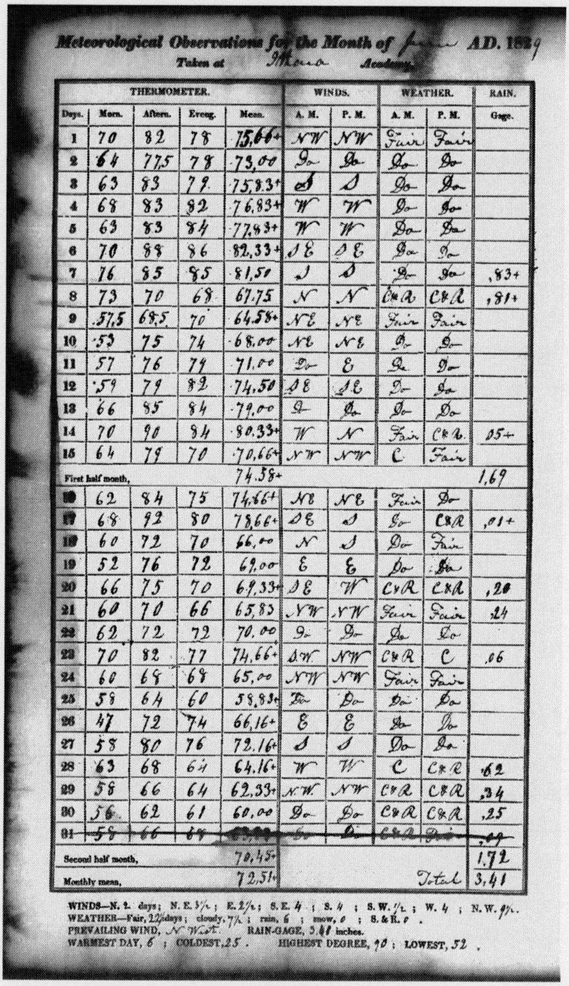
Meteorological Observations for the Month of June AD. 1829
Taken at Albany Academy.

THERMOMETER.					WINDS.		WEATHER.		RAIN.
Days.	Morn.	Aftern.	Eveng.	Mean.	A. M.	P. M.	A. M.	P. M.	Gage.
1	70	82	78	75,66+	NW	NW	Fair	Fair	
2	64	77,5	78	73,00	Do	Do	Do	Do	
3	63	83	79	75,83+	S	S	Do	Do	
4	68	83	82	76,83+	W	W	Do	Do	
5	63	83	84	77,83+	W	W	Do	Do	
6	70	88	86	82,33+	SE	SE	Do	Do	
7	76	85	85	81,50	S	S	Do	Do	,83+
8	73	70	68	67.75	N	N	C&R	C&R	,81+
9	57,5	68,5	70	64.58+	NE	NE	Fair	Fair	
10	53	75	74	68,00	NE	NE	Do	Do	
11	57	76	79	71,00	Do	E	Do	Do	
12	59	79	82	74,50	SE	SE	Do	Do	
13	66	85	84	79,00	Do	Do	Do	Do	
14	70	90	84	80,33+	W	N	Fair	C&R.	05+
15	64	79	70	70,66+	NW	NW	C	Fair	
First half month,				74.38+					1,69
[illegible]	62	84	75	74,66+	NE	NE	Fair	Do	
17	68	92	80	78,66+	SE	S	Do	C&R	,01+
[illegible]	60	72	70	66,00	N	S	Do	Fair	
19	52	76	72	69,00	E	E	Do	Do	
20	66	75	70	69,33+	SE	W	C&R	C&R	,20
21	60	70	66	65,83	NW	NW	Fair	Fair	,24
22	62	72	72	70,00	Do	Do	Do	Do	
23	70	82	77	74,66+	SW.	NW	C&R	C	,06
24	60	68	68	65,00	NW	NW	Fair	Fair	
25	58	64	60	58,83+	Do	Do	Do	Do	
26	47	72	74	66,16+	E	E	Do	Do	
27	58	80	76	72.16+	S	S	Do	Do	
28	63	68	64	64.16+	W	W	C	C&R	,62
29	58	66	64	62,33+	N.W.	NW	C&R	C&R	,34
30	56	62	61	60,00	Do	Do	C&R	C&R	,25
31	~~58~~	~~66~~	~~64~~	~~[illegible]~~	~~[illegible]~~	~~Do~~	~~C&R~~	~~Fair~~	~~,09~~
Second half month,				70,45+					1,72
Monthly mean,				72,51+				Total	3,41

WINDS—N. 2 days; N. E. 3½; E. 2½; S. E. 4; S. 4; S. W. ½; W. 4; N. W. 9½.
WEATHER—Fair, 22½ days; cloudy, 7½; rain, 6; snow, 0; S. & R. 0.
PREVAILING WIND, N. West. RAIN-GAGE, 3,41 inches.
WARMEST DAY, 6; COLDEST, 25. HIGHEST DEGREE, 90; LOWEST, 52.

FIGURE I.1. Meteorological data form, Albany Academy, 1829. From *Reports of Meteorological Observations Made at Academies, 1826–1859*. Courtesy of New York State Archives, series no. A0346-78.

tory of the "weather blank," a meteorological form, conceptualized over this century-long period, shows the persistence and evolution of formal numerical schema as well as distinct categorical arrangements for weather across time. As this book reveals, streamlining and codifying meteorological data collection passed between many institutions, from the Albany Academy (figure I.1) in upstate New York, to the Smithsonian Institution (figure I.2), the Commerce Department, the Department of Agriculture, the Department of War (figure I.3), and the Weather Bureau itself (figure I.4). Data collection began with volunteer communities in the nineteenth century and grew into a formalized, employee-based sys-

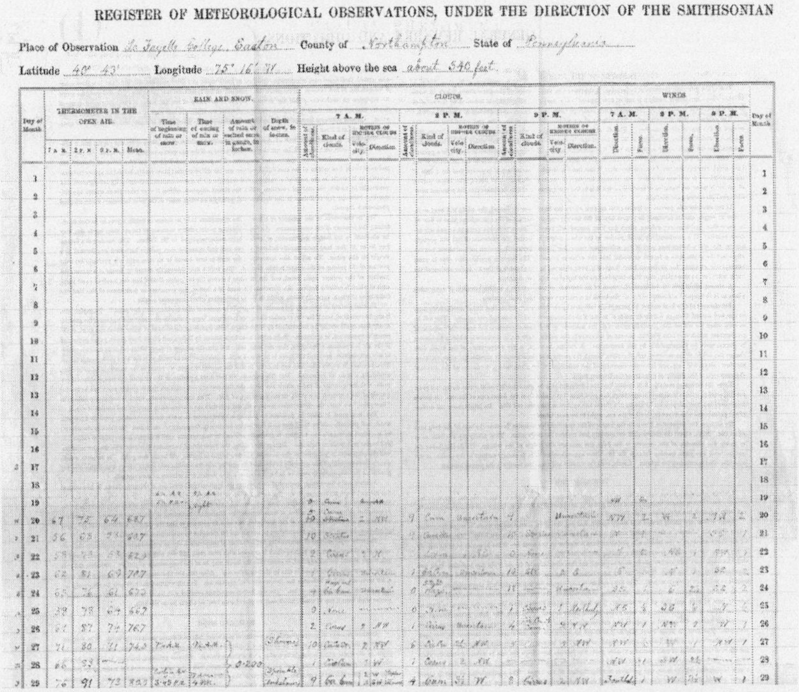

FIGURE I.2. Meteorological data form, n.d., likely 1850s or 1860s. Courtesy of Lafayette College Meteorological Records. Special Collections and College Archives, Skillman Library, Lafayette College.

tem in the twentieth century, which complemented but did not replace volunteer efforts.

Throughout this book, I use *data* as a singular and a plural. When I use *data* in the singular form, I am referring to the concept. When I use *data* in the plural form, I am referring to masses of observations collected by weather observers and remote sensing technologies. I refer to settler weather data cultures and settler environmental consciousness through this work. Both concepts have particular meanings. Settler weather data cultures are composed of material textual, graphic, or infographic practices associated with data collection, collation, reduction, and dissemination. When I refer to environmental consciousness, I am referring to the ways of thinking, relating, and feeling that grow out of that culture. Throughout the book, I call attention to the different technical regimes that made data collection possible, but this book is about data predom-

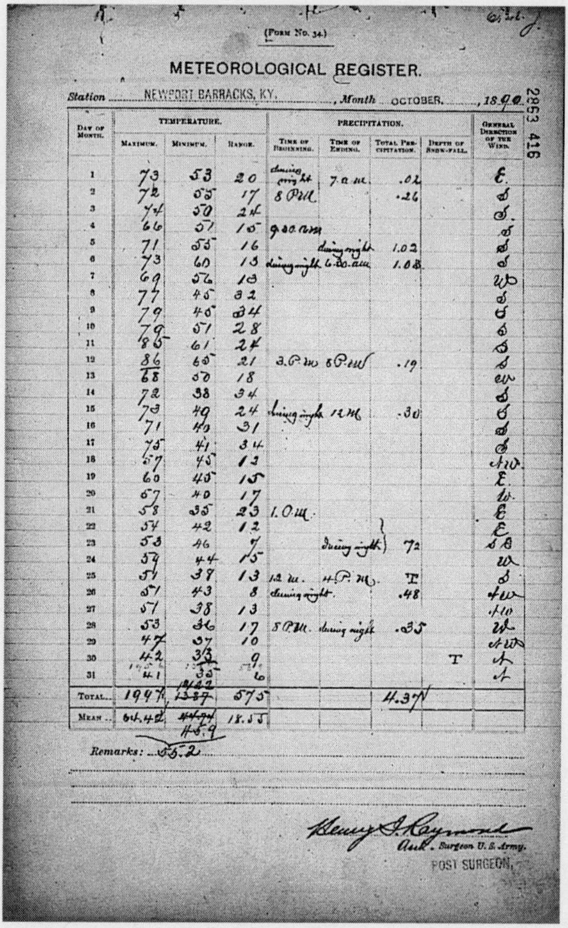

(Form No. 34.)

METEOROLOGICAL REGISTER.

Station NEWPORT BARRACKS, KY., Month OCTOBER, 1890. 2863 416

Day of Month.	Temperature. Maximum.	Minimum.	Range.	Precipitation. Time of Beginning.	Time of Ending.	Total Precipitation.	Depth of Snow-fall.	General Direction of the Wind.
1	73	53	20	during night	7 a.m.	.02		E.
2	72	55	17	8 P.M.		.26		S
3	74	50	24					S.
4	66	51	15	9.30 a.m.				S
5	71	55	16		during night	1.02		S
6	73	60	13	during night	6.30 a.m.	1.08		S
7	69	56	13					W
8	77	45	32					S
9	79	45	34					S
10	79	51	28					S
11	85	61	24					S
12	86	65	21	3 P.M.	8 P.M.	.19		S
13	68	50	18					W
14	72	38	34					S
15	73	49	24	during night	12 M.	.30		S
16	71	40	31					S
17	75	41	34					S
18	57	45	12					NW.
19	60	45	15					E.
20	57	40	17					W.
21	58	35	23	1 P.M.				E
22	54	42	12					E
23	53	46	7		during night	72		SE
24	59	44	15					W
25	51	38	13	12 M.	4 P.M.	T		S
26	51	43	8	during night.		.48		NW
27	51	38	13					NW
28	53	36	17	8 P.M.	during night	.35		W
29	47	37	10					NW
30	42	33	9				T	N
31	41	35	6					N
Total	1947	1387 / 1422	575			4.37		
Mean	64.42	44.74 / 45.9	18.55					

Remarks: 55.2

Henry G. Raymond
Asst. Surgeon U.S. Army.
POST SURGEON.

FIGURE I.3. US Army surgeon/sergeant observer, form no. 34, Newport Barracks, Kentucky, October 1890. Courtesy of Midwestern Regional Climate Center, accessed June 15, 2022, https://mrcc.purdue.edu/FORTS/forts_forms/AE04490/00899.JPG.

inantly and technology secondarily. In the later part of the book, I pay increased attention to these technical regimes, primarily because they deeply influence notions of data in popular consciousness. I also treat data, at times, as a form of "ecomedia," to borrow from Hester Blum.[70] Blum's attention to polar ecomedia, and its capacity to "register complex systems of ecological and environmental change," resonates with *Immeasurable Weather*'s study of data as media, especially in chapters 3 and 5.

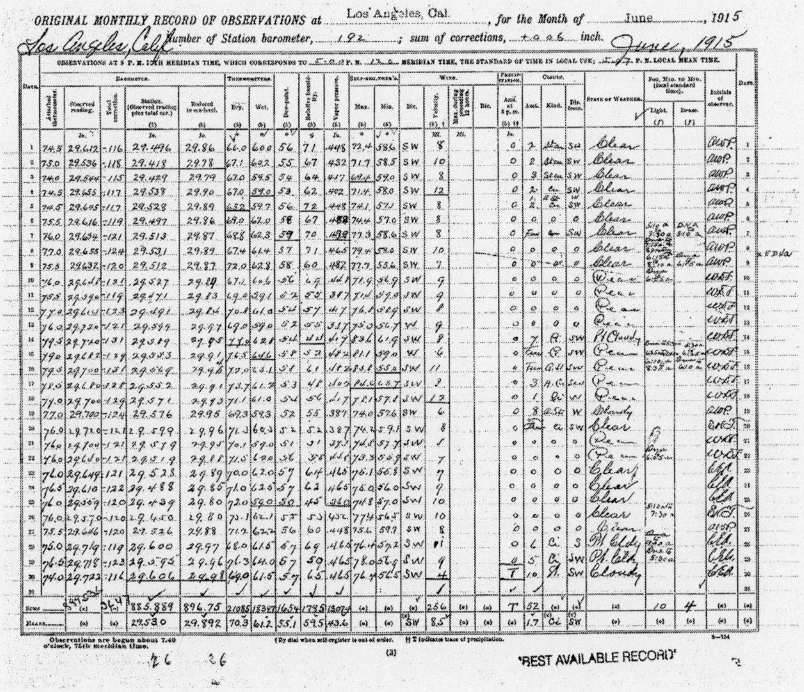

FIGURE I.4. Weather Bureau meteorological data form, Los Angeles, California, 1915. Courtesy of NOAA's National Centers for Environmental Information (NCEI), accessed June 2022, https://www.ncdc.noaa.gov/IPS/coop/coop.html.

Settlers upheld data as a tool for coming to know a world that did not belong to them, and yet one that they claimed. The more data settlers had, the more they thought they knew. But as this book will show, the systems they set up for knowing the very environments they inhabited had inherent faults: relationships of impact and reciprocity across biospheric communities were not part of the systemization. As data became more popular, it seemed to be the only way settlers could engage the world they had made. This book is about the worlds weather data creates and the legacies it leaves behind.

1

DREAMING DATA: LOCATING EARLY NINETEENTH-CENTURY WEATHER DATA

It had just begun to rain when Ithaca Academy professors S. Phinney and X. Haywood walked outside from their academy offices and stood in front of a hole they had dug in the ground. Occupying Cayuga land—as the academy had since the incorporation of Ithaca Academy in 1823—Phinney and Haywood began the process of measuring the rain.[1] They placed a hollow cylindrical vessel in the ground and waited for the rain to end; after the rain stopped, they measured in inches the amount of still water that had collected in the vessel and recorded this measurement in a book of twenty-four folio pages, foolscap size. Phinney and Haywood repeated this process every day that April—measuring rainfall but also temperature, wind, and sometimes cloud cover—until the entire month had been measured and recorded. The rain made it a busy month—shad, current, plum, and cherry all blossomed within a week of each other, and the barn swallows made their first appearance the very last day of April. The professors would note all of this, too, in their observation books.[2] The year was 1827, and across the occupied lands of New

York State, settler meteorological networks expanded their tendrils. With thirty-four stations that extended as far north as Canandaigua, Ontario, and as far south as Jamaica, Queens, professors and their students used folio books to design their monthly forms and yearly reports, which included monthly averages of temperature, records of wind direction, and rain totals. The reports also included miscellaneous observations—the first daffodil bloom, the Hudson free of ice.

Equally important as what was included in these reports was the language that compilers of these reports used to describe themselves. "First ice seen by the observer—surface of the ground quite frozen," wrote the Reverend John C. Rudd at the Auburn Academy in October 1827.[3]

Rudd's decision to refer to himself in the third person as "the observer"—rather than as a reverend or schoolteacher, the professions that employed him—speaks to the eagerness with which early-century settler communities wished to see themselves as participating in a science culture. As Rudd's language suggests, meteorological observers (or those who volunteered their time in pursuit of grassroots meteorological science) would emerge as a distinct category across the early to mid-nineteenth century. And with this network of statewide observers came a set of observing practices that were outlined, revised, reimagined, and enforced through a set of discursive formats, from instruction manuals to forms and annual reports. As Phinney, Haywood, and Rudd sought to define themselves as observers—and practiced a form of observation and collection that was so central to observer culture—early-century observers were also creating something else: archives of observations that would eventually become meteorological data. Over the course of the early century, the concept of data took conceptual, linguistic, and practical shape in settler meteorological culture.

In 1837, when the Franklin Institute's committee on meteorology remarked that meteorological observations are "made for the purpose of obtaining data," it crystallized a connection that had been brewing across early-century meteorological communities in the United States and abroad.[4] Observation was a basis for data, but it was not the total sum. Data was directed toward particular ends, as Joseph Henry—meteorological compiler for the New York State Regents' meteorological network—would suggest in 1833. But to what ends, precisely? This chapter explores how settlers understood the weather through numericity. In order to obtain numerical observations, they relied on land occupation and eventual theft. Settlers used meteorological observation—and, later, "data"—as a way to visualize territory and create an environmental consciousness grounded in the relationality of numbers for populations that had little historical experience with the environments they occupied. Throughout all this, settlers depended on the graphical form of the table for storing and processing new material known as "data"; they used graphical forms to manage weather observations, conceptualize data itself, and visualize a settler colonial nation through the natural laws and scientific truths such data allowed for.

The historical and geopolitical colonial science scene of the nineteenth century included robust imperial meteorological observation networks, maritime climate science, and a popular interest in weather prediction. Regarding the British and maritime contexts, Lorraine Daston has elaborated on the role of meteorological data collection and organization within the Early Modern Royal Society and the culture at large. Building on Daston, Katharine Anderson explores the turn to prediction from data collection in nineteenth-century Britain, as well as how data collected in imperial regions advanced British science and facilitated colonial control.[5] Additionally, as Jan Golinski outlines, British populations studied the weather with enthusiasm, keeping journals and diaries that were part of print culture and public life throughout the Enlightenment.[6]

As systematic weather observation gained popularity across settler and imperial nations, leaders in the burgeoning field of weather science attempted to standardize measurement techniques, from instrument type to calibration and instructions for collection.[7] From isothermal mapping to the development of international metrics and standards for weather measurement, the international weather science scene leveraged accumulated weather data observations to build an international climate and weather science based in, as Martin Mahony has argued, an "imperial meteorology" across the eighteenth and nineteenth centuries.[8] Based in ship logs and international weather station data, Alexander von Humboldt's and Edmond Halley's work attests to the ways that maritime data collection, charting, and mapping were closely aligned with British and US imperialism.[9] Humboldt's map of isothermal lines shifted understandings of climate across the second half of the nineteenth century, increasing spatial awareness of weather and atmospheric processes.[10]

Narrowing the scope down to the US settler state of the nineteenth century, this chapter considers the role of systemic weather observation via internal colonialism across the US settler state on the heels of the Indian Removal Act of 1830.[11] While this chapter does not consider white weather observers from southern states who enslaved Black populations, the work that eastern weather data observers did in the early part of the century laid the foundation for white southern observers to simultaneously practice weather data collection and enslavement later in the century, thus extending the operation of internal colonialism. The internal colonialism of the eastern part of the continent manifested on a continental scale—encouraging and bolstering a "nation"—by the middle to late nineteenth century. Additionally, meteorological data collection was supported by and fed "settler atmospherics," in the words of Kristen Simmons, the "normative and necessary violences found in settlement—accruing, adapting, and constricting indigenous and black life in the U.S. settler state."[12] As the story of data collection continues across two centuries, we will see how both internal colonialism and settler atmospherics bolster and support weather data science.

In the US settler state, observers not only used tables, the language of observation, and weather data to enact territorial expansion but also used these things to create settler narratives around environmental knowledge. Conceptualizing narratives of data would allow for settlers to build a national meteorological observation project that scaled up Phinney and Haywood's regional processes. These narratives encouraged a particular type of environmental relation and also powerfully articulated the view of a settler nation. The goal of a national data project relied on standardizing observation and understanding collections of observations as data.

As plans for a national project brewed through the efforts of regional leaders like James Espy and Joseph Henry, data—as a notion—was both conceptualized for institutions and publics and positioned as a public good. Yet the very nature of who was counted and valued in this public was at stake across the century. The sudden availability of regional weather data produced an insatiable national appetite for weather and climate knowledge among settlers. As more and more data became available—as more observations *afforded* data—climate shifted from signifying local or regional conditions to signifying a multiregional system. While the word *climate* was meaningful in a variety of ways prior to the 1850s—namely, as a means for recognizing hyperlocal weather conditions across time—the emergence of meteorological data as a category allowed for the rise of a notion of climate as a national and intergenerational condition. Thus, the emergence of data was linked at its outset to calls for public archives of climate knowledge. Just as a popular concept of data emerged, so, too, did a notion of climate.

These three processes—the way in which settler observers and institutions managed the practice of weather observation, the way institutions and institutes assigned the language of data to collections of observations, and, finally, how they joined data discourses with calls for public data as part of a nation-building project—illustrate how data was conceptualized in the US settler state through the social life of settler science. More precisely, weather observers and institution officials refined the contours of data through debates about visual display and, specifically, tables and related graphics. Tables would leave their mark on the environmental imaginations of institution officials and observers themselves, fundamentally altering popular understandings of weather and solidifying settler environmental consciousness through the numericity and compartmentalization of the table. Data was integral to the evolution of environmental thought, "integral" to the "social order" nineteenth-century weather communities would inhabit.[13] From academy professionals to weather observers and enthusiasts, from those producing instructions for generating tables to the tables themselves, environmental data laid the groundwork for nation-building.

MANAGING OBSERVATION

In January 1829, Phinney and Haywood sent all of their 1828 observations to Joseph Henry and T. Romeyn Beck at the Albany Academy. Henry and Beck were responsible for organizing and calculating the observations from Ithaca, and thirty-three additional academy offices, into a set of meteorological abstracts that would be published in the *Annual Report of the Regents of the University of the State of New-York Made to the Legislature*. By 1828, roughly forty-nine academies from across the state of New York were involved in daily weather observation.[14] Four years before Phinney and Haywood sent their observations to Albany, on March 1, 1825, then vice-chancellor of the University of the State of New York, Simeon De Witt, met with the Board of Regents to propose the project of statewide meteorological observation. De Witt's resolution was passed, and statewide observation commenced.[15] Taking daily, monthly, and yearly weather observations across New York State required many things, among the most important being settlement or continuous occupation of Indigenous lands. Creating one of the first statewide networks for weather data collection, collation, and publication, as well as managing the material and graphical output of observations themselves, hinged on white settler land possession, and this was something Simeon De Witt knew a lot about.

Settlers had been occupying the main tract of what is now known as New York up until the Revolutionary War and began squatting across the homelands of the Haudenosaunee by 1775.[16] The Haudenosaunee were and are an alliance of Native nations—including the Cayugas, Mohawks, Oneidas, Onondagas, and Senecas—that reside in New York and present-day Ontario.[17] In his actions as surveyor general for the Continental army during the American Revolution and the state of New York from 1784 to 1834, Simeon De Witt apportioned the homelands of the Haudenosaunee to settlers for permanent occupation. In coordination with his cousin, Moses De Witt, who was a land speculator, Simeon surveyed and mapped the way for white land possession in late eighteenth-century and early nineteenth-century New York.[18] From 1784 to 1835, De Witt was a key figure in apportioning and visualizing settler claims to lands, working with army captains who trespassed on Haudenosaunee Confederacy lands to survey "the Lands apportioned to the Troops of the State."[19] These surveys formed the basis of white settler land claims and paved the way for the settlement of meteorological observers like Phinney and Haywood.

De Witt helped to build the state's settler infrastructure, which allowed for the creation of a statewide meteorological network.[20] When in 1825 De Witt proposed that the academies begin to measure the weather in newly settled territory, he had already apportioned Cayuga, Mohawk, Oneida, Onondaga, and Seneca homelands to white settlers through land survey, speculation, and the legal apparatus of the New York State Surveyor General's Office. De Witt's proposal for data collection hinged on and was part of Indigenous land dispossession and settler dominance within the region. With that dominance in hand, De Witt proposed that each academy of the state incorporated by the board "be furnished with a thermometer and pluviometer or rain gauge, the expense of which will be paid out of the funds of the Regents."[21] The New York State Regents appointed a committee to "prescribe rules for observation" and to provide instruments to observers across the state. John Lansing and John Grieg—both lawyers—served on the rules committee. De Witt also served on the committee.[22]

Over the course of the 1830s and 1840s, the committee created a statewide meteorological network. Statewide observation opened the temporal and geographic range of weather display; it visualized settler territory through settler science yet bounded settler territory through quantified weather knowledge. Meteorology also created a specific environmental consciousness bound to numericity and tabular display. This consciousness was the latticework of occupation, solidifying what land surveying proposed—that settlers could claim Land. They sought to measure, study, and archive Land, turning Land into a resource to be studied, measured, and used for settler world-building. As a type of world-building violence, weather observation not only literally engaged unceded territory but also actively laid claim to it through a settler science founded on dispossession, theft, and genocide.

The statewide meteorological network distributed funds for scientific instruments, such as a thermometer and the rain gauge, to state academies. Though academies were asked to utilize the same set of instruments, from the common rain gauge to the thermometer, diverse types of instruments were now available by order through medical, farmer, and shipman catalogs, so observers sometimes procured their own. In addition to distributing instruments and instrument funds, the meteorological committee wrote and rewrote rules for weather observation and weather table construction several times throughout a twenty-year period: revised instructions were amended to the original set in 1825, and the 1825 instructions were reproduced with attached revisions in 1836, 1841, and 1845.[23] The revised instructions were a direct result of the difficulties in standardization the regents encountered when they attempted to compile and calculate observations for their annual publications of statewide data. The revisions that the regents made to the instructions for observation show that observation was uniquely tied to the visual form and the graphical possibility of the ordered, meteorological table.[24] And yet, observers struggled with order, often inventing their own rules for the visual display of weather. This frustrated De Witt and his committee because it hindered the process of developing a unified settler environmental data program. Weather observation—named data, later in the century—was key to centering settler pasts and outlining settler futures.

As De Witt, Lansing, and Grieg's instructions explained, each academy observer was expected to keep a yearly "journal" of "meteorological observations." Each month's tabular observations—taken every day, in the "morning," "afternoon," and "evening" for temperature, and once a day for categories of "winds," "weather," and "rain"—were to fill a single page in the yearly journal.[25] Observers were asked to follow a form appended to the end of the instructions. The form depicted the graphical organization of daily, monthly, and yearly registration of observations.

Directions for instrument use and placement were also detailed and lengthy, guiding observers in instrument positioning. The instructions also included a detailed summary describing how observers should handle and engage with the thermometer and rain gauge. Special attention was given to describing standards of objectivity for each of the instruments. Rain gauges were to be kept "remote of all elevated structures to a distance at least equal to their height." Thermometers were to be kept "in a situation where there is free circulation of the air, and where it cannot be affected either by the direct or reflected rays of the sun."[26]

The quality of observations sent to the Albany Academy for review, reduction, and eventual publication varied widely. Observers at Prattsburgh in 1826 neglected to record their thermometer readings throughout the month of August. Two notes are included next to Prattsburgh's incomplete August table, and both of them read: "Weather very warm." Other observers failed to note their locations on the tops of their folio sheets and books, while others decided to use the right-hand side of their tables for qualitative notes and averages for the month rather than utilizing the right-hand page of their folio book for this purpose, as De Witt, Lansing, and Grieg had requested. In Jamaica, Queens, Pierpont Potter was producing observations double the length of those produced in Prattsburgh. As the instructions requested, each monthly table contained three daily temperature readings ("morning," "afternoon," and "evening") and a temperature mean for each day, as well as an "A.M." and "P.M." "winds" and "weather" reading and a singular "rain" reading. In 1826, Potter devoted the left-hand side to numerical readings and the right side of the page to extensive notes on blooming schedules as well as related phenomena: "June 1, drought continues alarming, June 3, heavy thunderstorm in the night, June 10, hay harvest commencing" (figures 1.1–1.4). On June 19, Potter composed a half-page entry on a heavy rain that nearly broke his rain gauge.[27]

While Potter wrestled with describing the challenging elements of surviving a New York summer, other observers struggled with the organization of the table and the tabular transcription of their observations. As much as the weather tables were organizing a set of quantitative realities about weather in the form of temperature, wind, and rain measurements, they were also logs of qualitative experience with the weather and observers' experiences of the routine of measurement itself. For observers, qualitative explanation and addenda were not separate from the quantitative. Here, and later in the nineteenth century, tables are filled with numer-

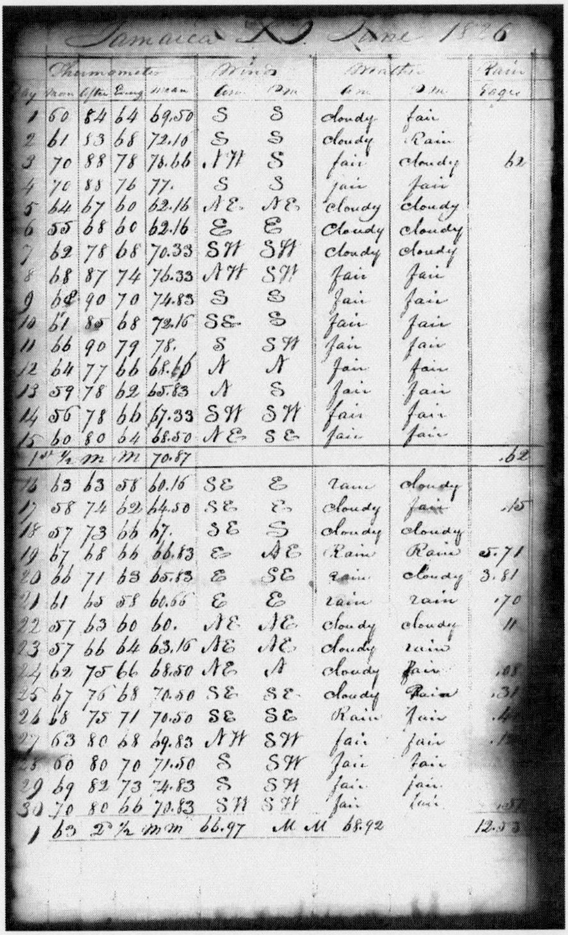

FIGURE 1.1. Pierpont Potter's New York State meteorological data form, 1826. From *Reports of Meteorological Observations Made at Academies, 1826–1859.* Courtesy of New York State Archives, series no. A0346-78.

ical readings and crowded with explanations of weather ephemera that do not fit into the layout of the table itself. Why did Potter include on his July 1826 return "Immense swarms of mosquitos such as is scarcely remembered by any of the inhabitants"?[28] Or, as on August 10 of that year, "Cool delightful weather"? Here, the record has a storytelling role to play. The stories observers tell—the questions they pose—enact place-based belonging and perform memory.

By 1836, the regents were fed up with the variety of observation tables they had been receiving throughout the first eleven years of the project. Incomplete observations were leading to incomplete portraits of the state's weather. In their edition of "Instructions from the Regents of the

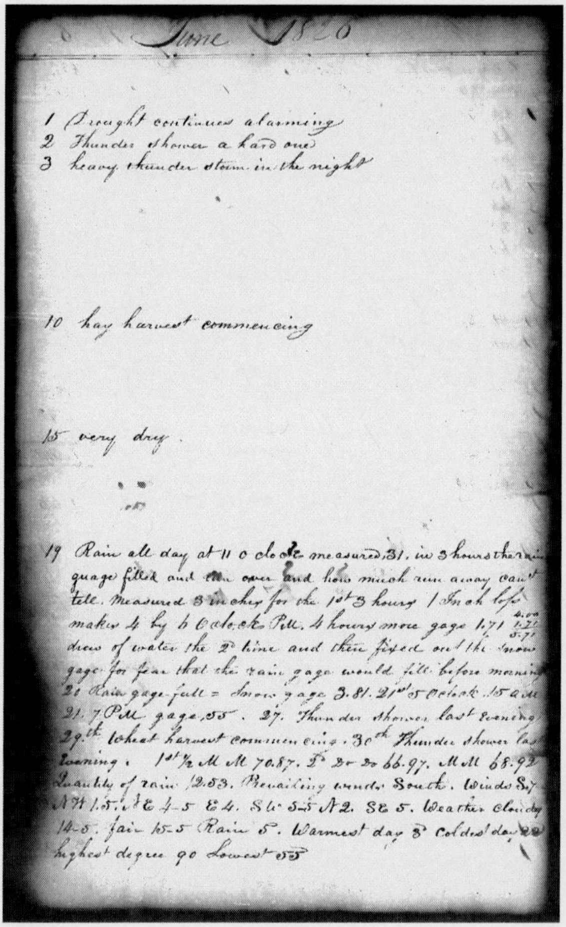
June 1826

1 Drought continues alarming
2 Thunder shower a hard one
3 heavy thunder storm in the night

10 hay harvest commencing

15 very dry

19 Rain all day at 11 o clock measured .31. in 3 hours the rain
guage filled and run over and how much run away can't
tell. Measured 3 inches for the 1st 3 hours 1 Inch lost
makes 4 by 6 O clock PM. 4 hours more gage 1.71 4.00 1.71 5.71
drew of water the 2d time and then fixed out the snow
gage for fear that the rain gage would fill before morning
21 Rain gage full = snow gage 3.81. 21st 5 O clock .15 AM
21. 7 PM gage .55. 27. Thunder shower last evening
29th Wheat harvest commencing. 30th Thunder shower last
evening. 1st ½ M M 70.87. 2d do do 66.97. M M 68.92
Quantity of rain 12.53. Prevailing winds South. Winds S 7
NW 1.5. NE 4-5 E 4. SW 5-5 N 2. SE 5. Weather cloudy
14-5. fair 15-5 Rain 5. Warmest day 8 Coldest day 22
highest degree 90 Lowest 55

FIGURE 1.2. Pierpont Potter's New York State meteorological data form, 1826. From *Reports of Meteorological Observations Made at Academies, 1826–1859*. Courtesy of New York State Archives, series no. A0346-78.

University" for that year, they discussed the state of the meteorological returns at length: "The meteorological reports from some academies are so deficient, notwithstanding all the instructions which have been heretofore given on the subject, that the Secretary finds it necessary to be more particular in his remarks than he has heretofore been."[29] In his second edition of instructions, De Witt first emphasized deficiency in form. Though observers were advised to "strictly and literally" follow the form of registering observations, many of them were not doing so. Frustrated with a lack of uniformity, De Witt described additional defects of the meteorological returns—columns left blank, averages uncalculated—stating that "such errors cannot be hereafter tolerated."[30]

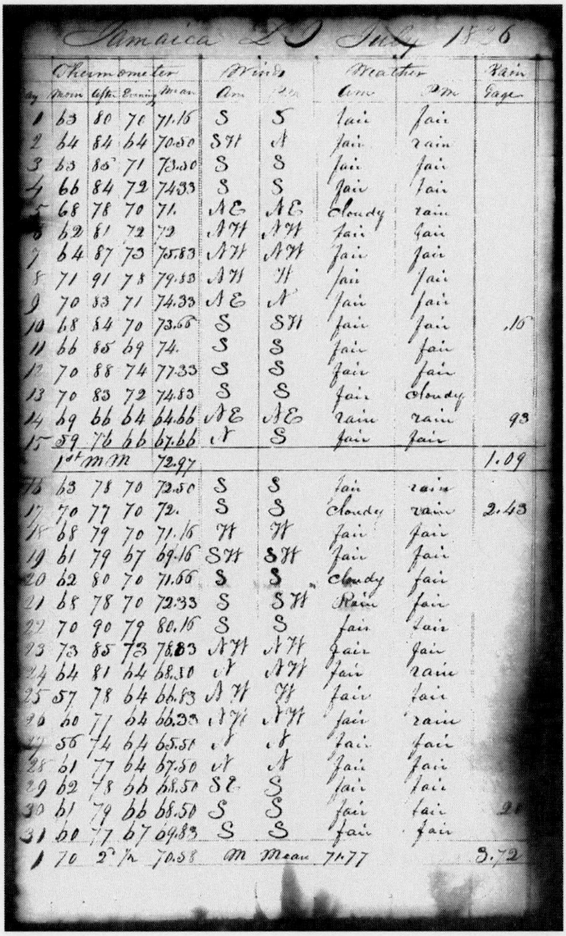

FIGURE 1.3. Pierpont Potter's New York State meteorological data form, 1826. From *Reports of Meteorological Observations Made at Academies, 1826–1859.* Courtesy of New York State Archives, series no. A0346-78.

The edited instructions reflected a very real anxiety around uniformity and completeness: tables were problematic for institutions when they presented incomplete observation or defied standard form.

Alongside form and completeness, the committee was also worried about the accuracy of human-gathered data. In their 1825 instructions, De Witt, Lansing, and Grieg attached a clause to their meteorological resolution that in addition to the existing regulations that "entitle the academies to the dividends of their public fund" in the form of the instruments that they were given free of charge, "it will be considered necessary that they keep an exact record of observations made with the thermometer & rain gauges, with which they shall be furnished." Observers were

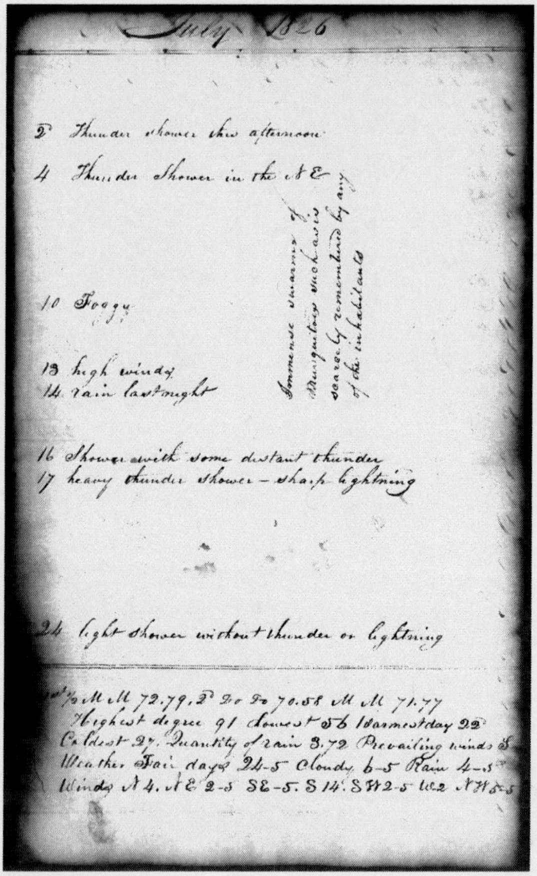

July 1826

3 Thunder shower this afternoon

4 Thunder Shower in the N E

10 Foggy

13 high winds
14 rain last night

16 Shower with some distant thunder
17 heavy thunder shower — sharp lightning

24 light shower without thunder or lightning

Immense swarms of Musquitoes such as is scarcely remembered by any of the inhabitants

½ M M 72.79. 3 Do Do 70.58 M M 71.77
Highest degree 91 Lowest 56 Warmest day 22
Coldest 27. Quantity of rain 3.72 Prevailing winds S
Weather Fair days 24-5 Cloudy 6-5 Rain 4-5
Winds N 4. NE 2-5 SE-5. S 14. SW 2-5 W 2 NW 5-5

FIGURE 1.4. Pierpont Potter's New York State meteorological data form, 1826. From *Reports of Meteorological Observations Made at Academies, 1826–1859*. Courtesy of New York State Archives, series no. A0346-78.

urged to "give correct register" of their observations.[31] Here, the committee focused on "exact[ness]" and "correct[ness]" of observations, illustrating a core issue for the regents and this project: the accuracy of distant observations performed by instruments and hands that were not their own. As weather data collection moves on throughout the early nineteenth century, the question of trustworthy data—correct, truthful, and exact—was something that each institution and government program would have to figure out on its own.

For New York State, this meant that no report would be accepted if it was not signed by the observer. Attached to the meteorological instructions were two sample forms—the "register" that observers were expected to follow, and an opening form that required the observer to sign beneath the following statement: "I [blank] of said Academy, do herby certify, that according to the best of my knowledge and belief, all and singular the meteorological observations, as registered in the following tables, have been correctly made, and truly registered by (Signed,)."[32] This signing off of the data recorded and often reduced (in terms of monthly and yearly averages) by observers was no small act. It was a meaningful moment in which the regents recognized the potential risks of misregistered and "untrue" data. If this kind of data were present in a form, they could track that recorder down easily. They could also enforce the data codes that they wished to be part of their recording regimes by asking observers to sign off on each data sheet—a signature that would verify not only that the observer had read the directions but that they had also adhered to the stated guidelines.

And yet, as sample tables and revised instructions show, observers often signed their names without adhering to the guidelines. The supposedly simple transcription of measurable weather phenomena into quantified reports was not simple after all. Additional challenges included the fact that observers were creating "errors" or "defects" by failing to complete their registers. Graphical forms such as the table were a staging ground for moments of graphical violation, and the discourse surrounding weather table construction was a place for managing weather display in its newly numerical form. The discourse that surrounded table construction was a form of disciplining observers, a way to articulate and enforce codes and conduct. The interplay of instructions, regulations, and the different forms the table might take all suggest that visual displays of observational data—and the environmental consciousnesses these displays would encourage—were far from clear throughout the 1830s and 1840s. As hard as institution officials might try to mandate forms for recording—as their goal was to discipline observers and standardize knowledge—observers would continue throughout this early period to use the table in experimental ways.

The experimental nature of building a settler environmental consciousness is best illustrated by the work of James Coffin, who was acting as the principal of the Ogdensburg Academy in the 1830s and began working with the regents' statewide network in 1837. That year, he sent a complete annual report to the regents that included not only his monthly and yearly folio tables, yearly averages, and extended miscellaneous phenomena but also several pages of notes reflecting on the relationship between the instruments used to measure the weather and the observations rendered. Coffin described the differences in rain gauge construction, wind observation, barometric observation, as well as a combined thermometer, barometer, and wind vane instrument and hygrometer instrument for measuring.

Since the mid-1830s, Coffin observed the weather, kept observation tables, and experimented with different modes of graphical layout for observations. Coffin published a series of weather tables in his *Meteorological Register and Scientific Journal*, a publication he designed, wrote, and issued in Ogdensburg, New York, in January 1839, and which lasted for a single printing. Coffin had high hopes for his publication and wrote in the inaugural and only issue that the *Register* "would be published monthly, each number containing eight pages. . . . It [the *Register*] will be a leading object of this paper to collect and collate these [New York State] observations; to trace, and illustrate when necessary by plates, their connections, or to furnish data which will enable others to do so and to show the order and succession of atmospheric phenomena throughout the state." Coffin understood weather data as an essential public good, noting that published data can "illustrate" weather realities and "enable" "others" to understand weather. He also believed that his Ogdensburg, New York, *Register* would eventually publish data from across the entire state. Though this would never come to pass, Coffin's aims were admirable and show that observers themselves were interested in weather table construction as well as public weather data.

Much like the regents' instructions, Coffin's *Register* centers a settler environmental consciousness that values numericity and categorical separation and display. But unlike the regents' articulation, Coffin is deeply concerned with visualizing regions across time through instrument use. The meteorological "problems" Coffin outlines in the first four pages of his publication range from how to find the mean temperature of a place to determining the "force of the wind." The solutions to these problems are often directly related to instrument use. For example, to solve problem "2d.—To determine the pressure of the atmosphere," an observer should "use a barometer" and correct the observation to a Fahrenheit standard. Coffin includes directions for anemometer use and measuring the depth of dew. The *Register* was a published account of regionally specific observations and an unofficial instruction book for amateur observers. All of this explanation was in the service of standardization, and standardization, itself, was in service of "illustrat[ing]" the settler state: when tabulated, observations—according to Coffin—would "show the order and succession of atmospheric phenomena throughout the state."

Coffin's work intrigued the New York State Regents, who in 1837 published an excerpt, devoting nine pages in their annual report to Coffin's methods and findings, including several tables of annual results from Ogdensburg. They remarked that "Mr. Coffin has extended his observations far beyond what is required by the Regents, and it is barely justice to notice his extraordinary labors."[33] Coffin's notes about instrument use and table formation were not simply extraordinarily detailed, extending far beyond requirements; they were, in fact, oppositional to the techniques of the New York State Regents. Different from the regents' rain gauge procedure, Coffin was weighing his water rather than utilizing a measuring stick. Coffin was also using a self-registering wind vane and reducing the readings to a single observation through trigonometrical calculations, whereas the regents had recommended simply counting the number of times the wind vane indicators point toward the compass. Coffin also had different observational and notational schematics for recording cloudiness, humidity and he had combined the barometer, thermometer, and wind vane into a single instrument. This allowed Coffin to track wind and rising temperature at the same time, the results of which he would reduce to a "single force" to "determine the point of maximum tendency to elevate the temperature." This meant that in a "single process the point of minimum temperature and of maximum and minimum pressure are determined."[34]

Some portions of Coffin's *Register* were not shared or discussed in the annual report, specifically the woodcut that Coffin included at the end of the *Register* (figure 1.5). The woodcut attempted to display both numerical and sensorial information from Ogdensburg side by side. Coffin writes that it is meant to "illustrate the connection between atmospheric pressure, the face of the sky, and storms made in a month's time" in 1838. Coffin's woodcut represented not an error or a defect within the statewide project but rather an "extraordinary labor." While the New York State network would not, in fact, shift its instrument use, observation recording, or observation calculation procedure based on Coffin's publication in 1838, the regents would begin to publish additional papers by Coffin in its annual reports, a report that was read by the regents themselves and the academies they served. In 1840, they published a set of Coffin's tables titled "Tables Showing the Prevailing Directions of Winds in the State of New York Thus Incorporating Observer Knowledge into the State-wide Effort."[35] And, in 1844, they published one of the longest mean temperature records for the region to date, "Average mean temperature of each

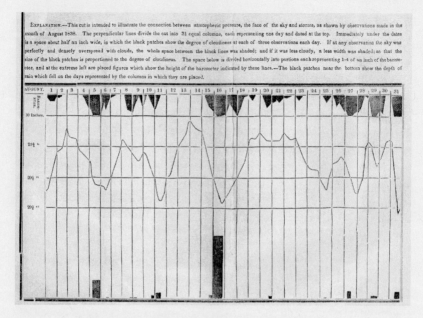

FIGURE 1.5. Woodcut from James H. Coffin's *Meteorological Register and Scientific Journal*, 1839. Ogdensburg, NY. Courtesy of NOAA Central Library.

day of the year, at Albany, N.Y., for a period of 20 years," which was calculated solely by Coffin based on the observations collected by the University of the State of New York since 1825.

In his preface to the table (which appeared in the regents' annual report as a letter to T. Romeyn Beck, who had served as the secretary for the Board of Regents of the university as well as the head meteorological compiler for the New York meteorological reports), Coffin remarks: "Dear Sir, The collection of meteorological observations taken under the direction of the Regents of the University of the State of New York, embrace so many localities, and now extend through a period of so many years, that they furnish valuable data for determining the law by which the temperature of different seasons of the year varies." When Coffin had access to roughly twenty years of meteorological reports and observations, his calculations produced something wholly new—data. Coffin uses data to describe the sum of settler environmental knowledge across space and time. Under the rubric of observation—and, soon, data—settler weather observations built a world of settler weather laws.

On September 6, 1833, Joseph Henry stood over a dead body in Morris-town, New Jersey. Henry had made the trip to Morristown from Prince-ton, New Jersey, where in 1832 he had begun a new position as professor of natural philosophy at the College of New Jersey, now Princeton University. Henry was traveling with another Princeton professor, Samuel Ladd Howell, and three of his students to perform galvanic experiments on the body of a murdered man—Antoine Le Blanc.[36] But something strange happened on the way to Morristown. Halfway there, in the town on Somerville, Henry and his companions noticed a faint light in the sky. The faint light grew and "gradually assumed the form of an arch extending around the north towards the east and west." It was a "beautiful aurora," Henry wrote in his notebook, which "embrac[ed] more than 70 degrees of the circle of the horizon." As Henry's description of the aurora ends in his notebook, he scribbles a drawing of the parallel beams that formed "as it were[,] the rudiments of an arch."[37]

While the notebook entry goes on to discuss the series of galvanic experiments conducted on the body of Le Blanc, the aurora left such an impression on Henry that he sought observations from several people in the eastern region, hoping eventually to make sense of the auro-ra's height and its relation to the magnetism of the earth. Henry asked for observations about the aurora from Peter Bullion, John Turnball—a Scottish friend who was traveling in Canada during the event—as well as William Kelly of Quebec. Across all these conversations and requests for observations related to the September 6 aurora, Henry and his cor-respondents began to use a new term to describe collections of observa-tions specifically aligned toward solving the meteorological quandary of the aurora's height and its relationship to the magnetism of the earth. This term was *data*.

Referencing the aurora observations that he had sent to Henry earlier that year, Peter Bullion wrote to Henry on December 19, 1833: "Have you not been able to make anything of our blundering data. [*sic*] Next time I shall be more accurate."[38] And in a letter to William Kelly in December of that year, Henry wrote, "I regret that I did not sooner direct my attention to the subject of the height of the aurora as I might have collected data sufficient to solve the problem while in Albany since I had the direction in connection with Dr. T. R. Beck of the Meteorological Reports made by the several academies of the State of N.Y. to the Legislature."[39] Both Henry and Bullion use the language of data to talk about aurora observations. Whereas Bullion locates his observations as a form of "data"—though "blundered"—Henry understands data as a mass of observations, collected into reports. In both cases, Henry and Bullion utilize the language of data in connection to solving the problem of the aurora's height.[40]

Compiling multiregional settler data was key to the story and enactment of settler environmental dominance. Henry was no stranger to requesting and compiling meteorological observations. In 1828, two years after being appointed a professor of mathematics and natural philosophy at the Albany Academy, Henry composed (along with T. Romeyn Beck) the *Abstract of the Returns of Meteorological Observations Made to the Regents of the University* for the year 1827. Henry would continue as the codirector of the meteorological observation network up until his departure from Albany to Princeton in November 1832. While in Albany, though, he learned the nuts and bolts of calculating regional observations and organizing statewide abstracts. He also realized the opportunities inherent in having access to statewide observations in tabular, standardized form. In a letter to John Maclean on June 28, 1832, Henry wrote: "I am considerably interested and have hoped at some future time to deduce many facts from it [the regents' report] of importance to the science of meteorology."[41] Henry's explanation of his interest to "deduce many facts" from the meteorological reports suggests his understanding that the reports themselves were of extraordinary value and that they could be used to arrive at meteorological truths. In this endeavor for truth, Henry and others would define masses of observations as data.

Connected to his interest in meteorology and magnetism, auroras were a long-standing interest of Henry's and were directly connected to the utilization of "data" as a keyword in this period. Between 1827 and 1832, Henry was engaged in various aurora pursuits. In 1827, Henry wrote to Lewis C. Beck—the brother of T. Romeyn Beck—describing the "fantastic flirtations of Albany auroras," and on April 25, 1831, at the Albany Academy, Henry shared his account of the relationship between an observed aurora borealis in Albany on April 19 and the earth's magnetism through observation of a needle.[42] Henry sought observations about this particular aurora from Parker Cleveland on February 14, 1832, requesting "any record of an aurora borealis having been observed at Brunswick or any other part of your state on the evening of the 19th of April 1831."[43]

Why were auroras among the first meteorological phenomena to be understood through the language of data in Henry's writing? The reason had to with Henry's suspicions and conclusions about the nature of auroras, namely, that they "cannot be classed among the ordinary local meteorological phenomena" but are rather connected to the "general physical principles of the globe."[44] The particularly general nature of the aurora as a meteorological phenomenon allowed the language of data to emerge. In their globality, Henry sought masses of observations—a curated and calculated selection—that would become data. Data were not reducible to a singular observation but were composed of many observations that spread across time and geography. When, in 1939, Henry wrote to his friend Alexander Dallas Bache—who himself had collected and aided in aurora observation gathering throughout the 1830s with Henry—and requested that Bache "collect the data for a map of isothermal lines of this country," he was utilizing this same notion of data as an intertemporal and interspatial product.[45] Though the data would be used in service of isothermal representation rather than solving aurora height, the meaning of data here is one that exceeds the local scale of observation. The use value of meteorological observations—whether these observations document ground or sky conditions—turned them into data, and utilizing weather observations was key to the emergence of data. Utilization emphasized extracting knowledge from environments for the purposes of nation-building.

Data universalized local and geographically disperse environmental experiences. Under the heading of "data," disperse observations were bound under the spatial rubric of the state, as well as the moving boundaries of the nineteenth-century settler colonial nation. The universal positioning of data by settler scientists is precisely what made the term so useful for building an environmental consciousness in the early to middle part of the century. Data was a powerful mode of scientific capture that, in the US context, asserted that environments could be known and understood through instrument measurement, tabulation, and categorization. The use of the word *data* in settler meteorological communities was bolstered by the weather table's emergence as the singular visual form to display weather observations throughout the late 1820s and 1830s. As observations were amassed and "embrace[d] so many localities" and "extend[ed] through a period of so many years," those same observations—stored in the table—turned into data for Bullion, Henry, and others across the period. On the local level, weather observation tables—"registers" and "journals"—made available conditions across a single location as well as a single day or set of days. On the statewide level, tables highlighted trends and averages across locations and time, while also reconstituting temporal experiences of weather (orienting temperature, wind speed, and even rainfall around the spatial notion of a state). Statewide tables displayed a numerically bounded territory through the language of data.

Other emerging meteorological projects outside of New York State were forming their own meteorological observation systems that, combined with those of New York State, would not only impact popular notions of data moving forward but also solidify the historical relationship between weather data and settler territorial imaginations.[46] Under the leadership of James Espy—known as the "storm king"—the Franklin Institute began, in 1831, to publish tabular records of the weather that were the result of Espy's call for instrument-based recording at the institute.[47] In his essay "On the Importance of Hygrometric Observations in Meteorology, and the Means of Making Them with Accuracy" (1830), published in the *Journal of the Franklin Institute*, Espy writes, "The object I have in view is to induce the Institute to use their influence to have meteorological journals kept in different parts of our wide extended country, on the same plan, and published in the journal of the Institute monthly. Many persons, for their own private gratification, are now keeping very imperfect meteorological journals who would feel a tenfold interest in the subject if they were informed by the Institute how to keep a much more perfect journal, with no more trouble, and at the same time contribute greatly to the public good."[48] Espy wanted institutional support for a meteorological project that would utilize the help of volunteers who already were "keeping . . . journals" across "our wide extended country." In 1831, his "wide extended country" was a question, not a reality, and the dream for a meteorological network as a "wide extended country" was articulated on the heels of the Indian Removal Act of 1830. Land surveyors and speculators—like Simeon De Witt and his cousin Moses— stole Haudenosaunee homelands for the sake of white settlement.[49] This allowed them to create the "wide extended country" on which Espy planned to build a scientific network.

Across New York and Pennsylvania, and throughout the eighteenth and nineteenth centuries, settlers stole land and resources for the enactment of meteorological science. They did this through a network of settler institutional practices, from land speculation and survey to squatting. From 1737 to the early 1830s and beyond, Pennsylvania, New Jersey, and Delaware settlers defrauded the Lenape of Lenapehoking through many decades of false treaties and continuous land dispossessions that forced their relocation from present-day Pennsylvania to various areas throughout the US. Beginning in the 1830s, the United States "made eighty-six treaties with twenty-six Indigenous nations between New York and Mississippi, all of them forcing land sessions, including removals."[50] Settler scientific institutions around the New England and northeastern region were applying observation and environmental data production across "*our* wide extended country" (emphasis added), producing weather data and knowledge and creating an environmental imagination of a state—and later country—literally bound by weather data.

Espy's Franklin Institute tables differ in scope, scale, and content from those of Henry and Beck. Rather than publish monthly averages, the Franklin Institute's table lists daily readings of each item, with three readings of temperature per day. The table notes "Thermometer," "Barometer," "Wind," "Water Fallen in Rain and Snow," and "State of the Weather and Remarks." It is important to note that while Henry and Beck's tables utilized numerical standards for wind reading, Espy's tables list wind direction and "Force," which ranges from "Blustering" to "Calm." The other interesting column is "Remarks," which is entirely occupied with descriptions of the sky, from "clear" to "cloudy" to "snow squall." Finally, there is no listing of location to indicate where these weather readings come from, though the "Committee on Publications" promises that "particulars in relation to the locality of the observations, to the instruments used, &c. will appear in our next number."[51]

Much like De Witt's instructions for New York State observers, central to the Franklin Institute's vision of its meteorological project is the disciplining of observers through observational instruction. In 1834, the Joint Committee of the American Philosophical Society and the Franklin Institute published the "Circular in Relation to Meteorological Observation" and set forth the language of sensorial meteorological observation. This positioned observation as a mode of information gathering—a pathway to data—and also linked observation to the sensorial capacity of the settler body. As Espy instructed in the *Annual Report of the Franklin Institute* for 1834, "The plan that we recommend in observing slow moving clouds, is to keep the head steady in one place, with the top of a chimney, or some distant fixed object, between the eye and any remarkable point of a cloud, until this point shall have moved so far from behind the object, as to leave no doubt of its direction."[52] Here, Espy talks through the practice of observation by instrumentalizing the settler body.

By 1837, observation was linguistically deployed to define meteorological data. In the January issue of the *Journal of the Franklin Institute*, Captain Alfred Mordecai—a military observer stationed at Fort Arsenal, Pennsylvania, for the "extended country" meteorological project—writes, "The observations which form the subject of this communication were made at the suggestion of a friend, (Professor A. D. Bache,) who is a member of the American Philosophical Society and the Franklin Institute, for the purpose of obtaining data for determining the laws of diurnal variations of temperature."[53] Mordecai's language positions data as something to be used "for determining" meteorological laws; observations of the extended country were made "for the purpose of obtaining data." As a notion, data was understood as a mass of observations, and observations themselves were founded on the sensorial capacities of the observer.

Data is positioned, here, through the user's ability to trace difference across temporal scales. Mordecai uses his observations to chart mean temperature across a six-month period.[54] He continues on to say that "to obtain further data for ascertaining the hours of daily mean temperature, it is not necessary to embrace the whole day in our observations, for the results in Table 1, furnish evidence of a striking fact, which enables us to determine, the mean temperature of the day, by observations of the maximum and minimum."[55] Mordecai ends his report by saying that he will not pursue his "investigation" any further at this time and hopes "some persons favorably situated might supply us with additional data for confirming or correcting the laws of variation of temperature, which have been laid down in this paper."[56] Mordecai's report makes several usages of data clear. First, data rises out of observation and is used for "ascertaining." Second, data and observation will, when organized into a table, "furnish evidence of a striking fact."[57] Third, data is neither true nor false in itself—nor is it private—but is rather held in common and used to confirm or correct meteorological laws.

Much like auroras, storms were interspatial events that traveled across locations and connected distant territories through a singular weather phenomenon (figure 1.6).[58] In its "Third Report of the Joint Committee on Meteorology," which follows Mordecai's report, the committee explains that its two previous reports "gave an account of nine storms," and "two very remarkable storms have occurred since." The report goes on to explain that "immediately after the termination of these storms, the committee addressed circulars to a great many individuals throughout the United States, requesting an account of them, particularly with regard

The storm which commenced on the night of the 19th June is not less remarkable than the one just described. It seems also to have traveled S. or S. W., for on the night of the 19th, and the next day, the rain was greatest at Silver Lake, in the northern part of Pennsylvania, and on the night of the 20th, and the next day, it was greatest at Baltimore, and it did not reach Hawsburg, Rappahannock county, Va., till the 21st, when it rained very heavily, the wind having veered round to the N. E. At Staunton, Va., it was clear all day the 21st and 22d, till evening, and the N. E. storm did not set in till the 24th.

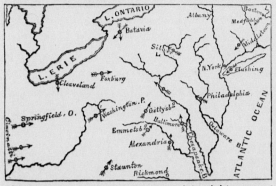

In the wood cut we have given the direction of the wind by means of arrows at all the places from which we have any accounts. On the 20th of June, there fell nearly three inches of rain, including that of the preceding night, at Silver Lake. Much more rain fell here than at any of the surrounding places of observation. There was but little at Foxsburg, Gettysburg, and Harrisburg, and it did not commence at Baltimore till 7h. 30m. P. M., nor at Emmettsburg till 4 P. M., nor at Philadelphia till 5h. 10m. P. M.

By casting the eye on the wood-cut, it will be seen at a glance that the wind blew on all sides towards the point of greatest rain.

But what is not a little remarkable, the wind was much more violent all round the borders of the storm, than near the centre.* At Harrisburg and Gettysburg it was gentle, and at Emmettsburg it was tempestuous. It is true it did not begin to be tempestuous till 4 P. M., and at Philadelphia it did not begin to blow very hard at the surface of the earth, till about 11 A. M. At 7 A. M. it was very calm, though the low clouds were flying with great velocity from the S. E., exactly towards the region of the great rain.

It is also worthy of particular remark, that the wind at Emmettsburg was N. W. on the 21st, blowing towards a region where it rained that morning and the preceding night, so much as to do great mischief, and even as near to Emmettsburg as Baltimore, it rained in that time 1-7 inches.

* It has since been ascertained that during the day of the 20th, the wind at Silver Lake was so variable that it was difficult to know from what point it prevailed most.

FIGURE 1.6. Woodcut depicting the direction of the wind. From "Third Report of the Committee on Meteorology," *Journal of the Franklin Institute* 19 (1837): 19.

to the rain, and the strength and direction of the wind."[59] The committee then used these observations to prove that "air moves inward toward the center of great rains." To prove this meteorological truth, the committee used "acknowledged data" arrived at "by calculation."[60] Just as Mordecai had positioned data as a material that rises out of observation—that is "obtained" by it—and can be used to "ascertain," the committee on meteorology makes a similar reference. Data are calculated.

In total, there are nineteen instances of the word *data* across the 1837 report, fourteen of them meteorological in nature. This included references in two reprints of scientific articles, "Of the Influence of Different Winds on the Heights of the Barometer" and "Simultaneous Meteorology." In both cases, the language of data is less about outlining bodily procedures for weather observation (how one should turn their head, where they should fix their eyes), and more about what masses of observation afford the study of weather. The first article reads, "Till data sufficiently numerous be obtained to enable us to attempt the solution of these various meteorological problems, I shall here present to the reader the results of two series of very accurate observations."[61] In direct opposition to two singular observations, data is positioned as a mass directed at a particular end—determining the height of an aurora or the path of a tornado. In "Simultaneous Observations," we see: "For the sake of affording [meteorological] data . . . this Committee would most earnestly recommend . . . quaternary division of the 24 hours."[62] Observations afford data.

In the early portion of the nineteenth century, meteorological data was a sign of settler possession: meteorological data culture operated within a discourse of territorial expansion and ownership, and the culture circulated a set of environmental relationships founded on continuous occupation of land for settler scientific gain. Within meteorological communities of the 1830s, the language of data emerged as a way to give shape to stolen territory. Settler meteorologists amassed and collected distant observations to answer particular queries and fostered a data culture founded on theft of land, knowledge, and life. They instrumentalized the settler body to address colonial environmental problems and propose solutions. As data comes to signify a collection of observations that exceeds a singular locale (and comes to supplant the language of observation) it begins to embrace so many localities that observers begin to see the outlines of "climate."

DATA NATION

"The number of observations cannot be too great," wrote James Espy in his circular "The Friends of Science" in 1842. Espy would echo these words in 1843 in his *First Report on Meteorology to the Surgeon General of the United States.* He had been hired by the War Department in August of that year to report annually on the "territory embracing the observations" in which "simultaneous observations [are] about to be established."[63] Espy wrote that forms would be sent and journals kept in Bermuda, Canada, Newfoundland, Nova Scotia, the West Indies, and the Azores, as well as on vessels sailing in the Atlantic or the Gulf of Mexico, and at all the lighthouses and naval stations. He encouraged the 103 colleges in the US settler states to request and fill out daily forms and wrote that "more observers are much wanted, especially . . . in the 'far west,' northwest, and north, and in the ocean between this and Europe." Espy imagined that if "such a system of wide-spread simultaneous observations, continued long enough," "mankind" could "compare with each other, not merely the storms of different seasons, but the storms of different years!" Here, he imagined an empire of data collection, expanding the geographic range and temporal bounds of meteorological observation not only across the continent but also across parts of the globe. Through three different actors who had already been building settler meteorology networks through the language of data in the early part of the century—James Espy, Joseph Henry, and James Coffin—this imagination of a national territory bound by weather observation would become a reality over the next ten years. And as it became a reality, a new practice would inform meteorological culture in the US settler states: the practice of bounding a nation through climate knowledge and climate data archiving.

In 1837, Espy asked Joseph Henry of the Albany Academy, as well as representatives of societies and colleges, to "write to some member of Congress to advocate" for a bill that would help build "meteorological observations throughout the United States." The bill, which was presented to Congress by John Sergeant—a member of the House of Representatives from Pennsylvania—did not pass, but the effort shows Espy's energies for expanding the geographic range of meteorological observations. Espy also appealed to Congress and the public in an 1837 newspaper article, which requested that communities keep "journal[s] of the weather" to be sent to Espy "monthly in Philadelphia."[64] Finally, Espy was involved in heated debates throughout the 1830s and 1840s in the "American storm controversy," which included scientific characters such as William Redfield and Robert Hare.[65] He shared his theories on storms by participating in the lyceum movement.[66] Throughout the storm controversy, as well as throughout the popularization of meteorological thinking brought about through public lectures, Joseph Henry and James Espy remained closely aligned—Henry was not only familiar with the work Espy was doing with the surgeon general throughout the 1840s but also participated in the lyceum circuit, maintained letter correspondence with

Espy throughout 1842, and, in 1845, was sent a copy of Espy's report to the surgeon general by James Forbes.[67]

In his commitment to data collection, and his public efforts more broadly across the 1830s, Espy positioned meteorological observation and data production as a national practice.[68] But this positioning shaped notions of "nation" itself, as well as shared understandings of environmental pasts and an environmental future that centered settler ideation. Espy's first report on meteorology submitted during his employment by the surgeon general was addressed "To the Friends of Science," a seemingly vast public he imagined as willing and able to participate in simultaneous observation. In explaining the process of observation and collection, he wrote that "all who send me a faithful account of the winds and weather will contribute to the great end in view. Editors of papers, who notice great storms, may be of much service. . . . Let none think their *mite* too insignificant to be thrown into this common treasury."[69] Simultaneous observation was a "common treasury," a collection of "faithful accounts" made by "all." Yet, the "all" of this apparent meteorological common was, to date, exclusively white, male, and settler.

Much like his sentiments when working as the chair of the Joint Committee in the 1830s, Espy articulated that making observations available to settler publics would contribute to the "good" of that public itself. In his essay "On the Importance of Hygrometric Observations in Meteorology, and the Means of Making Them with Accuracy," he wrote that a distributed network of meteorological observations that could be printed and made available to any interested party would "contribute greatly to the public good."[70] Although state and national borders were contested in this period, Espy bolstered a sense of the nation as an imagined community with defined borders and settled publics connected through environmental conditions such as the weather and climate. This imagined community would be realized through maps, designed by Espy and printed by the Surgeon General's Office.[71]

Although Espy sought meteorological observations from various locales that exceeded the continent, he predominantly organized his maps around the Eastern Seaboard and eastern inland. These maps helped to visualize these regions as settler states, and domains of settler environmental knowledge, furthered by Espy's promise to send "each of my regular correspondents" the report of maps.[72] Observers who received these maps would be able to review, utilize, and visualize the weather of the "United States"; in return, Espy would obtain enough data to fully develop his theories on storm dynamics.[73] Espy's very public work—both his desire to send meteorological reports, maps, and data to his correspondents and his visual display of meteorological observations on a cartographic plane that bound a supposedly settled nation—paved the way for an environmental consciousness among settlers that used data capture (and the land dispossessions such capture was built on) as its core relation to environment and land. This way of relating would gain national popularity among settler volunteer communities reporting to the Smithsonian Meteorological Project—a national science project with international aims—headed by Joseph Henry, from 1847 to 1870.

Espy's work throughout the 1830s and 1840s emphasizes the accessibility of settler and international colonial environmental data, an approach that harmonized with James Coffin's. In his *Meteorological Register and Scientific Journal* (1839) and his 1844 publication of twenty years of temperature data in the New York State Regents' reports, Coffin also expresses the necessity of accessible data. He writes that the journal will "collect and collate these observations; to trace, and illustrate when necessary by plates, their connections, or to furnish data which will enable others to do so and to show the order and succession of atmospheric phenomena throughout the state."[74] Coffin understood weather data as an essential good, noting that furnished data can "illustrate" weather realities and "enable" "others" to understand weather. In creating this essential good, Espy, Coffin, and Henry presented a nation bound together through weather systems and the data they were built on. Meteorological observation and the proliferation of weather data more broadly in the early nineteenth century was an "enumerative strategy" that created a coherent colonial space grounded in environmental concerns, conditions, and geography within the settler colonial imagination.[75]

While Espy was busy gaining congressional support for meteorological observation and building a settler public hungry for data gathering, Joseph Henry forged a new meteorological relationship with James Coffin, one of great importance for the future of data in the US settler state that would last well into the 1870s. Like Espy, Coffin—the New York State observer and the creator of the *Meteorological Register*—advocated for weather data as a public settler good. Henry had seen Coffin's 1839 feature on the meteorology of Ogdensburg in the annual report of the regents of the University of New York and so wrote to him that year, enclosing a copy of "Contributions III: Electro-Dynamic Induction" with the inscription "To James H. Coffin, author of the interesting report on the Meteorology of Ogdensburg, just published in the abstracts of the Regents of the University of New York. With the respects of Joseph Henry."[76] Henry and Coffin would correspond throughout the 1840s, and these communications would lead to not only a plan for a national meteorological system—the Smithsonian Meteorological Project—with a "central point" for "matters of science" but also the publication of two meteorological data reports that enlarged conversation about data, climate, and climate variance in the US settler state.

In 1842, Coffin and Henry exchanged letters regarding astronomical tables and education, as well as the status of meteorological observations at Princeton University and the "phenomenon" of thunderstorms, a topic related to Espy's position in the storm controversy.[77] In 1844, they discussed an opening mathematics professorship at Lafayette College—a professorship Coffin would come to hold—and in July 1845, Coffin explained to Henry that he had been working on collecting observations and producing meteorological data as well as wind roses (a graphic that provides a view of wind speed and direction for a specific location) for a report on the winds of North America and the Atlantic Ocean to be presented to the American Association of Geologists and Naturalists. Coffin said that he found himself "most deficient in observations from the southern states," and he wondered if Henry could help him "enlarge" his list of potential observers for the South.[78]

A little over a year later, on December 3, 1846, Joseph Henry learned that he had been elected the first secretary of the Smithsonian Institution. Coffin wrote to Henry to congratulate him and to offer some advice about the "object & design" of the institution, writing that it might "serve as a medium through which to concentrate scientific observations in different parts of the country, & by its publications afford to the various observers, the benefit of each others [sic] labors."[79] Much like Espy, Coffin wanted national data—to afford scientific publications to various observers—as well as a "central point" for the "embodied knowledge of the country on matters of science." He saw that potential central point as the Smithsonian, much as he would like his idea to use Ogdensburg as a central point for meteorological observations of the state of New York.[80] Henry returned Coffin's letter a little more than a week later, writing that the plan of organization "will include your suggestions."

In 1847, James Coffin provided Joseph Henry with "a list of all persons, as far as they are known, who have heretofore been accustomed to make meteorological observations in North America," writing that the list "will be of much importance in our future investigations relative to this subject."[81] In 1848, after organizing this list, Henry sent circulars to his congressional colleagues, as well as to the meteorological communities that had been building around the nation since 1820, writing that interested parties should "signify their willingness" to participate in an "extended system of meteorological observation, particularly with reference to American storms, embracing as far as possible the surface of this continent" by writing to the head of the navy, James Espy, to request instructions and instruments.[82] Letters began rolling in from late 1848 to early 1849, and the observer count began to grow.[83] In 1849, the Smithsonian counted 14 volunteer observers across "this continent," a relatively small number for the task ahead.[84] By 1872, the meteorological project counted 536 observers, with a note that the same 14 observers or observer locations counted in 1849 had been with the project since its 1848 inception.[85] In the early years of Smithsonian development, from 1847 to 1849, Henry remained in contact with Coffin, traveling to Philadelphia to see Coffin present his "Report on the Winds of the Northern Hemisphere" at the 1848 meeting of the American Association for the Advancement of Science.[86] After this meeting, Henry wrote to Coffin about the possibility of presenting his wind manuscript in a series of publications called the *Smithsonian Contributions to Knowledge*. This agreement blossomed into a partnership lasting several decades in which Coffin would not only publish his wind manuscript with the Smithsonian but also use volunteer data collected after 1848 to enrich the manuscript itself. As Henry drew up his "Programme of Organization," he prioritized Coffin's suggestion for affording observers the benefits of each other's labor and Coffin's suggestion to position the Smithsonian as a central point for matters of science; he also designated that part of the yearly income would be spent on the study of meteorology. In conversation with Coffin, Espy, Bache, and others, Henry developed a plan of organization and presented it to the regents for adoption. The December 8, 1847, "Programme of Organization" read that "a [s]ystem of meteorological observations for solving the problem of American Storms" would be a "special object" of research at the Smithsonian, with funds annually set aside for its flourishing. Henry's system of meteorological observations—no doubt deeply influenced by the New York State meteorological effort and Espy's Joint Committee effort as well as his work with the surgeon general—would eventually become the National Weather Service.[87]

While the Smithsonian Meteorological Project is most clearly linked with the project of the US settler state, Henry and his contemporaries had international ambitions for meteorological data collection. In James Espy's meteorological circular of 1842, he announced his ambitious plan for "widespread simultaneous observations" that would extend across "the United States, Bermuda, West Indies, Azores, and the Canadas."[88] Henry also desired to create a similarly widespread system. In his 1849 Smithsonian Institution report, he remarked that "we are in a fair way of establishing a general system of meteorology, extending over a great portion of North America," with "prospects of procuring permanent observations from Bermuda, some of the West India islands, and from Central America."[89] Years later, in 1859, Joseph Henry would meet in Washington, DC, with the governor of the Bahama Islands, who "promised to procure" a "series of meteorological observations" for the Smithsonian Institution's network.[90] The Smithsonian also maintained relationships—though terse—with the Board of Education of Canada and the Hudson Bay Company regarding systems of meteorological observations.[91] As historian Kae Takarabe has shown, the Smithsonian Institution also maintained an "International Exchange Service" whereby the Smithsonian Meteorological Project "enabled Henry to introduce systematic meteorological observations to parts of Japan and to collate this data."[92]

After the inauguration of the Smithsonian Meteorological Project, the language of data increased across meteorological documents. In 1850, Henry wrote that "with observation . . . data will be furnished for important additions to our knowledge of meteorological phenomena."[93] In the *Annual Report of the Board of Regents of the Smithsonian Institution* for 1864, Henry wrote that observers belonging to the national meteorological project "contribute[d] their services without compensation. Their only reward is the satisfaction of co-operating with each other and the Institution in the effort to supply data and materials for investigation."[94] But beyond this increase, the word *data* was used to describe something that extended beyond a mere collection of weather observations. The Smithsonian's sixth annual report included "Report on Meteorology" written by the Committee on Meteorology of the American Association for the Promotion of Science. The committee members aimed to assess extended systems of meteorological observations, including Espy's work, now under the heading of the navy, as well as efforts at the Smithsonian. In their report, they called for thirty-nine new meteorological stations and wrote that with these new stations, they "expect to derive from systematic observations, extended over as much of our continent as is accessible to us, at stations selected in reference to the problems to be made out, a thorough knowledge of our climate in all its relations, and of its variations in the same and in different localities."[95] The study of "our" national climate "in all its relations" and "variations" across time and place became a priority. This priority was rooted in what the Smithsonian had begun referring to as "collection[s] of meteorological data."[96]

The relation between data and climate became even more precisely defined by 1855 when Joseph Henry writes that the meteorological project at the Smithsonian affords of "approximate data" "relative to the climate of the United States."[97] He also had the institution put out a call for "general phenomena of climate." This call included "date of greatest rise and lowest fall of water in large rivers" and "important annual staples" such as "general leafing" and fall of leaf in deciduous forests, as well as the "breaking up of ice in large rivers or bays."[98] By 1857, the Smithsonian announced it plans to publish an eight-year collection of what it referred to in various documents as "meteorological observations," "meteorological data," and "climate knowledge," explaining that "all the materials collected at the Institution are in the process of being arranged and bound into accessible volumes, with proper indices, to be used by all who may be desirous of making special investigations on any point relative to the climate of this country. . . . The materials which have been collected relative to the climate of the North American continent [include] the observations made under the direction of this institution since 1849." And it was none other than James Coffin—Ogdensburg meteorologist and Joseph Henry's friend—who would compile *Results of Meteorological Observations, Made under the Direction of the United States Patent Office and the Smithsonian Institution from the Year 1854 to 1859*, a six-year collection of weather data in the United States.[99] In his introduction to the collection, Henry wrote that the reductions of observations "furnish data of importance in studying the peculiar disturbances of the atmosphere to which the North American continent is subjected" and that "all the data accumulated in regard to it should be given to the world."[100] Echoing both Espy's and Coffin's calls for a nation of data, Henry released the most comprehensive collection of meteorological data to the settler public. His reason for doing so was to allow these publics to relate the "general laws of climate" through "condensed abstracts."[101]

The study of climate "profited" from newly available meteorological data.[102] Settlers had access to meteorological data and received encouragement to solve climate "problems" via this data. They shaped conversations, accessed institutional discourses, and received printings of meteorological data in exchange for their services. Settlers across North America furthered the work of state-building and nation-building through weather science and forged a new concept of data and national climate that would captivate generations of settlers to come. Most important, they built a data culture that was founded on conquest—chattel slavery and Native genocide, land theft, dominance, and control of biosphere life through measurement. *"Our* climate" was a settler organizing principle—much like *our* extended country was for Espy—that depended on and deepened conquest primarily because it assumed endless access to land.

The process of dreaming and building a data culture underscores how, as Liboiron has argued of pollution, "science always happens within land relations, and those relations are always specific to a place, even if you don't believe in Land."[103] On the logic of "separation," Liboiron continues, "The naturalization of separation allows the scientific logic of variables to make sense—variables are ways to treat elements of an environment as discrete, autonomous actors. . . . This isn't to say variables are 'wrong' or fundamentally colonial things (they don't automatically grant access to Land for colonial goals), but to point to how worldviews allow some things to make sense and act as truth at the expense of other things." The "at the expense of other things" is crucial here. For in constructing a data culture that prioritized discrete entities—and one might even say discreteness is essential to the concept of data—observers foreclosed other forms of relation with Land and self. The worldview of conquest disallowed these relations. Data culture attempted to know weather discretely, each element an "autonomous actor"—though many observers struggled with this when completing their weather data observation sheets—separate from the human and more-than-human relations that bound each place of measurement.[104]

2

GENDERING DATA:
THE WOMEN OF THE SMITHSONIAN
METEOROLOGICAL PROJECT

In 1858, Anna Bowen—a settler on Omaha (UmoNhoN) lands, now called Elkhorn City, Nebraska—began reporting as a volunteer meteorological observer for the Smithsonian Meteorological Project.[1] Bowen and her family came to Nebraska Territory in June 1857. Anna's father, John S. Bowen, was born in New York, and her mother, Annette, in Pennsylvania. In 1857, they settled on forty acres, built a house and stables, and started measuring the weather.[2] In her first month of observation, Bowen recorded thermometer and "face of the sky" data three times every day for the entire month.[3] Bowen was also interested in precipitation but lacked the instrument—a rain gauge—necessary for making these readings. Instead, she marked a line of rain on her returns for that year and sent it in, hoping that the makeshift measure would suffice. In the same letter, she requested a rain gauge from the Smithsonian Institution. Later that year, Bowen wrote on the bottom of her submitted register for the month, "The rain gauge promised us never came to hand—Can you send us one?" A note scribbled on Bowen's letter, which now resides in the Smithsonian archives, stands as a kind of response: "Wait a little longer."

It is not clear who read Bowen's note, or whether this scribbled response was ever sent back to Bowen herself. What is clear is that Bowen never received a rain gauge in 1858 or in 1859. In fact, throughout Bowen's entire volunteer relationship with the Smithsonian—one that would last until 1864—she would never receive the rain gauge she had requested.[4]

And yet, by 1860, three institution-identified male observers in Bowen's region of Nebraska Territory—James Allan, the Reverend William Hamilton, and John G. Rain—had access to and were actively utilizing rain gauges for the meteorological observations they were sending to the Smithsonian Institution. Both Allan and Rain were somewhat unreliable observers: Allan's returns were infrequent, as in 1861 when he returned only two monthly meteorological registers for the entire year, and though Rain had been furnished with a full set of instruments, he would only report back to the project from 1860 to 1861, a single year. Another Nebraska settler and meteorological volunteer, John Evans, who began reporting for the project one year after Bowen's requests for a rain gauge in 1859, also utilized a rain gauge for his Smithsonian weather measurements. Yet he, too, was an unreliable observer, reporting for only five out of twelve months in 1863. Anna Bowen, on the other hand, never missed a month of observation in the eight years she reported to the Smithsonian Meteorological Project.

Recently, scholars of information have pointed to gaps in feminist and women's studies of data labor. Jennifer Light and Mar Hicks offer untold histories of women computers and gender discrimination in the mid-twentieth century.[5] Catherine D'Ignazio and Lauren Klein approach this from a contemporary lens, advocating for a "data feminism" as a way of bringing bodies back into discussions about data collection and cultural data work, such as content moderation.[6] Yanni A. Loukissas argues for studying data through a "local lens," showing how data is "inseparable" from "social, material, and spatial conditions."[7] I show that there is historical precedent for the contemporary critique that D'Ignazio and Klein rightfully locate within the data sciences today, and I delve further back, historically, than Hicks and Light to offer a nineteenth-century example of women, data labor, and gender discrimination. The work I do here to bring women back to histories of data labor reassembles data history at the "local level," as Loukissas advocates, uncovering the social, material, and spatial conditions of mid-nineteenth-century data labor.

The nineteenth century was an emergent period for the professionalization of science in the US settler state, as Sally Gregory Kohlstedt, Kara

Swanson, George Daniels, and Nathan Reingold have shown.[8] But even more than a time of professionalization, this period was a revolutionary time for white women in particular, as Kohlstedt and Swanson elaborate. In this period, white women attempted to gain access to, and practice within, newly professionalizing scientific communities in ways that were not possible earlier in the century. More specifically, the emerging field of meteorology, which was developing through grassroots settler scientific practice across the "extended country," offered white women an opportunity to become a part of an extended scientific network open, at least on paper, to all "friends of science."[9] The Smithsonian Meteorological Project—headed by Joseph Henry, the first secretary of the Smithsonian Institution—was a correspondence network that relied on the postal service and, for this reason, allowed participation from men, women, and those who wished not to disclose their genders.

Historians of women's science have shown that while women were able to address this seemingly open scientific network, their efforts to become part of this community were met with resistance and sometimes erasure.[10] This chapter adds to that story, providing further evidence of this resistance and erasure within meteorological science, attending to the ways that "gender differences" were "used to rationalize power differentials and divisions of work in the home and the labor market."[11] Far from a historical anomaly, Anna Bowen's experience can be understood as representative of this historical moment. Bowen was not the only self-identified woman who requested, but struggled to receive, materials for volunteer observation as part of the Smithsonian Meteorological Project. Her story is one entry point into a complicated narrative about the ways that self-identified and self-disclosed women in the mid-nineteenth century became a part of a white male, settler colonial meteorological science. This weather observation, data collection, and data reduction network was a flexible one in that different women within it were offered different forms of engagement by white male officials. White women's exclusion and complicity worked to solidify this network under dominant patriarchal norms.

Patriarchal norms affected all self- and institution-identified women regardless of social class. Yet, intersectional factors—such as race, religious affiliation, and citizenship status—shaped access to settler science under patriarchal norms. Settler science—from natural philosophy to botany—was highly visible in private and public schools from 1830 to 1889.[12] Yet as Kabria Baumgartner elaborates in her book *In Pursuit of*

Knowledge: Black Women and Educational Activism in Antebellum America,
while public schools were ostensibly "open to all social classes," "access to
public high school, let alone achievement, depended largely on the socio-
economic status and racial background" of a family.[13] While it is true
that by 1840 the institution of slavery was "all but extinct" in the northern region, "white supremacy thrived nationwide, leaving free blacks in
a precarious position" where "custom" rather than "law" shaped access
to education.[14] Public and private schools in the antebellum North were
sites of exclusion and contestation, where African American women and
girls fought against the processes of racialization that shaped access to
and notions of education.

While African American girls were often denied the right to education
—protesting these inequalities while also creating alternative systems
for education—within the dominant white system, missionaries influenced Creek and Cherokee access to settler science in the early nineteenth century. Donald Fixico documents the multiple missionaries
who worked among southern tribes between 1805 and 1816 to establish schools in the Moravian, Presbyterian, Methodist, and Baptist traditions.[15] At least one mission school—the Episcopalians' Cass Lake
Mission School in Minnesota—reported weather data to the Smithsonian in the 1850s.[16] Devon Mihesuah has shown how training in botany
and farming was part of the curriculum for Cherokee girls at the Cherokee Female Seminary.[17]

The apparatus of weather data collection—communicating data
through the postal mail—made the misogyny of weather data collection
different than the misogyny of other scientific practices where men and
women shared physical space. The second section in this chapter documents this difference, while the third section elaborates how misogyny
functioned in physically shared space. Midcentury meteorological data
was deeply tied to the sensorial and emotive capacities of white men. By
surveying their letters, drawings, and data conversations, we can trace
the ways in which these male communities shaped meteorological data
between two distinct poles—data was a deeply sensorial, body-anchored
process that depended on facilities of the senses (from sight to sound
and often memory) and an instrument-based, numerical material. Male
observers sought to embrace data as haptic matter, while Smithsonian
Institution officials sought to stabilize data as a numerical, tabular entity.
When white women were admitted into this project, they were treated
in markedly different ways—as Bowen's case suggests—than their male

counterparts at the time. This was true in the field but also in Easton, Pennsylvania, where a group of fifteen women computers worked to copy and calculate meteorological data for the Smithsonian Meteorological Project. At Lafayette College, James Coffin—head mathematician for the Smithsonian Meteorological Project agreed in a written commitment to the Smithsonian's "incoming and accumulated observations" to utilize the labors of women public school teachers from Easton, Pennsylvania, to copy and calculate multiyear meteorological data.[18] The names of these women never made it to the Smithsonian or Patent Office billing books, nor did they receive special thanks in the final publication of materials.[19] The tale of two groups of women laboring for meteorological science—one in semipublic and the other in private, one performing the initial collection of meteorological data and the other tasked with its calculation—provides evidence of white women attempting to stake their claim to settler scientific work as early as 1847.

This narrative will travel from Washington, DC, to Easton, Pennsylvania, to Nebraska Territory and back again. The historical characters in this story include famous women, like Clara Barton, as well as women computers whose work shaped scientific practice in the mid-nineteenth century but whose names could not be recovered. Taken together, the named and nameless women volunteers, computers, and copyists who were part of this project helped to build the foundations of a settler colonial meteorological data collection and computation system in the United States. This chapter shows how women were both under- and over-utilized, incredibly visible and persistent volunteers and yet also hidden away performing copying and calculation work for the project. In the kitchens in Easton, Pennsylvania, and in the basement of the Patent Office, the fact that women were in physical proximity to men meant that their labors were cordoned off to semiprivate or hidden spaces; white women "engaged in scientific work" but "remained on the periphery of the scientific community" whether they were in the field performing meteorological data readings or in their kitchens or the Patent Office basement performing copying and calculating work.[20] White women data laborers were gendered through self- and institution identification. Sometimes, in their letters (and by their pens), data laborers noted themselves as "Miss" or "Mrs." Other times, institution logbooks and letters from institutions classified them through the language of "female" and "lady." Self- and institution assignment of masculinity worked similarly. Gendering data, and gendering data laborers, were complex exchanges between those reporting to institutions, institutions themselves, and preconceived notions of gender.

Although James Espy and Joseph Henry—creators of the meteorological project—promoted the vision that meteorological volunteer work required no special training, the skills associated with volunteer-based science were positioned by men as masculine traits. But the records show women performing this very same work across the period.[21] As was true in other domains where women attempted to claim their place—such as women attempting to enter clerkship and merchant work at time—there was nothing "inherently masculine" about any of these forms of labor other than that men had traditionally performed them.[22] Studying the letters and labors of the ladies of the Smithsonian Meteorological Project—from the field, to the kitchen, to the Patent Office basement—shows how weather data collection and calculation were socially embedded processes for self-identified white women at the time.

These materials also reveal the ways in which white women aided in the processes of settler science, often participating in and advancing the settler colonial logics—from land dispossession to enslavement—that came along with a nation-building meteorological science. Many of the midwestern women who reported as meteorological volunteers to the project were illegally occupying Indigenous lands, with the desire to settle, while collecting meteorological data; women and men in the antebellum South who were acting as weather observers were also slaveholders.[23] The very notion of a nation was shifting and contested in 1848; weather data collection helped to stabilize a notion of the nation as settled and white, though data collection itself took place in contested landscapes where only certain racialized and gendered bodies were able to take part in the making of weather data. The myth of a white and predominantly male settler nation traveled hand in hand with meteorological science, and while white women can at times appear to be the heroes of this story—staking their claim to scientific life—they are far more complicated (compromised and complicit) figures within this story.[24]

MAKING DATA

On the morning of June 24, 1852, Professor P. A. Chadbourne, volunteer weather observer in East Windsor Hill, Connecticut, captured two locusts—male and female—from a nearby field.[25] He placed both locusts, still breathing, into an envelope and mailed them to the Smithsonian

Institution's Meteorological Project office in Washington, DC, along with a letter that read: "I hope they will reach you alive. . . . Is there any particular observation that ought to be made regarding these? Will you please forward me more maps for observation on the Aurora?"[26] Unfortunately, we do not know whether the insects reached the Smithsonian alive, but we do know that the Smithsonian received the locusts along with the letter and that this letter—like many others containing insects, rocks, flowers, beehives, hail drawings, tornado maps, and thousands of weather data "blanks"—arrived at the meteorological office between 1847 and 1870, becoming part of the material beginnings of the Smithsonian Meteorological Project.

Perhaps it was Henry who opened the locust-laden letter in the balmy summer days of 1852, or Edward Foreman, whom Henry had hired as a meteorological clerk for the Smithsonian.[27] As was common practice when the Smithsonian received letters for the Meteorological Project, each letter was separated from the attached materials (or, in this case, the enclosed insects) and placed with other weekly and monthly correspondence. Among the correspondence from this particular June, Henry had received several requests for blanks (the standardized form produced by the Smithsonian—and graphically linked to the tabular forms described in the previous chapter—for recording temperature, air pressure, cloud cover, and wind direction) and an observer's report on the Albany earthquake of April 1852.

The enclosure of living locusts in an envelope destined for the Meteorological Project office in Washington, DC, suggests the diversity of "weather data" for male observers in the midcentury. For our Connecticut locust gather, weather was related to the migration and habitation of insects. For other observers, the pursuit of weather data would include the study of the shape and quality of eclipses, hail, and the composition of stone (figures 2.1–2.2).[28]

Male observers approached the study of weather and the collection of data with a certain categorical flexibility, utilizing weather as a meeting place for a variety of studies that had yet to formulate themselves as distinct fields in American scientific institutions, including entomology, upper atmospheric science, dendrology, and botany. Meteorology acted as a holding place or primer for many different emerging scientific fields.[29] As a holding place, it also came to stand for the utopianism of scientific progress. After learning of the institution's efforts, a male weather enthusiast from Trenton, New Jersey, wrote "[I] . . . was requested once to pub-

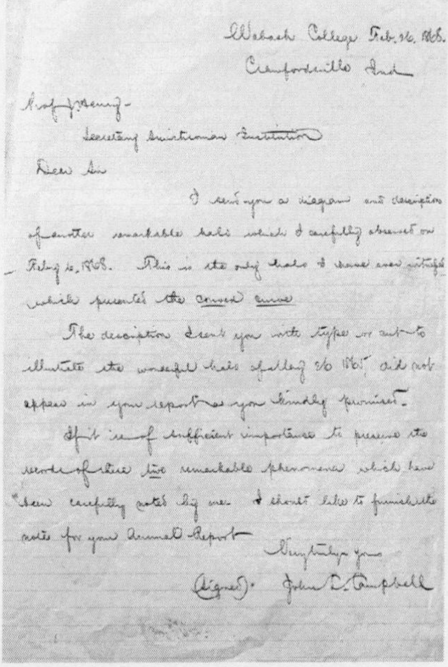

FIGURE 2.1. Letter describing a hailstone, 1868, John L. Campbell. Courtesy of the Smithsonian Institution Archives, Record Unit 60, box 7, folder 6: 2/26/1868, Smithsonian Institution, Meteorological Project, Records.

lish a report of my observations in the papers—this, I declined doing, but feeling that it will be duly appreciated by your institution, and encourage scientific investigation, have prompted the report to you."[30] Male observers were eager to send their own local recordings to the institution if it would "encourage scientific investigation" or "facilitate the progress of scientific research" (as another observer put it in 1852).[31] They were emboldened to send their own scientific materials to the institution for review.

Male observers understood that there was an exchange value attached to the environmental information they provided the institution in the form of meteorological observations, or blank forms. In a letter dated May 1, 1852, from Dr. Samuel K. Jennings sent from "the City of Austin, Texas," an observer enclosed three meteorological forms, submitting them to the institution for review, and wrote: "I send some small pebbles marked 1. 2. 3. 4. and a piece of white stone marked 5. Let me hear from you on the capacity of no. 5 to be applied to building purposes."[32] As is common across many letters, observers begin by furnishing the Smithsonian with the weather information that it had requested and often submit a request for additional blanks. This observer, like many others, wanted information of their own from the Smithsonian and was not too shy about asking for it. With eagerness and trust for the Meteorologi-

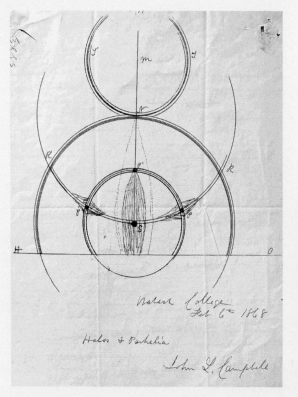

FIGURE 2.2. Drawing of a meteorological phenomenon, 1868, John L. Campbell. Courtesy of the Smithsonian Institution Archives, Record Unit 60, box 7, folder 6: 2/26/1868, Smithsonian Institution, Meteorological Project, Records.

cal Project and the scientific venture it represented, observers sent the Smithsonian material for "scientific" testing.

In July 1852, a male observer wrote to the Smithsonian Meteorological Project inquiring into the shape of hail: "A few months since, at 3 o'clock in the afternoon, we were visited with a very heavy hail storm, which lasted about 10 minutes, wind from the North and sky clouded all over. The dimensions of each hail stone and form was thus a perfect cone, the large end or base being solid ice, graduating to soft ice at the top. . . . Will you gratify me by informing me of the reason of it being thus shaped?"[33] The observer includes a drawing of the hailstone in this description, with the graphic interrupting the letter itself (figure 2.3).

In its linguistic and graphical presentation of hailstones, this letter shows the desire of male observers to contribute to the categorical flexibility of weather and weather data collection, as it extended beyond mere recording of temperature, wind, and precipitation. Both this and the Texas request reveal that male weather observers were constantly in the

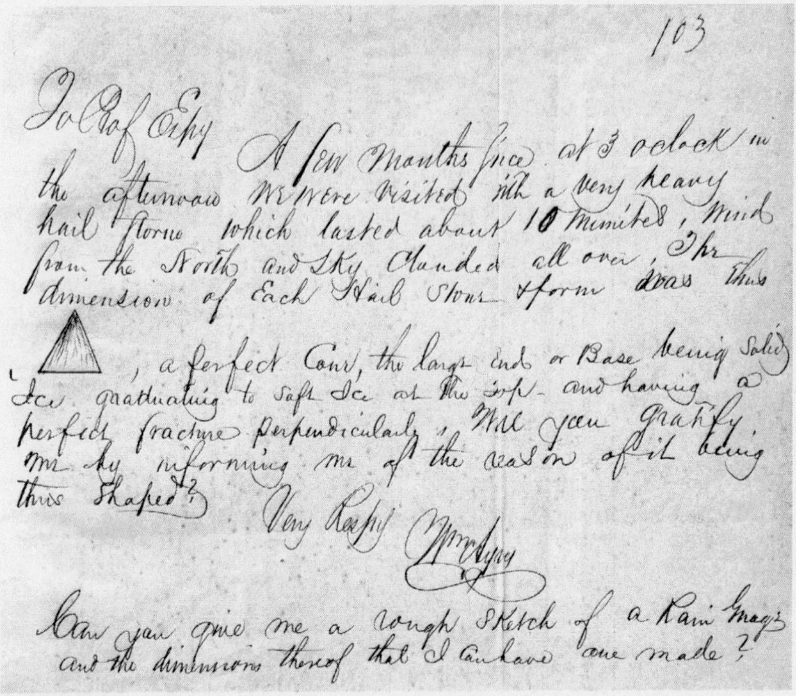

FIGURE 2.3. Letter from a volunteer observer, 1852, depicting hail. Courtesy of the Smithsonian Institution Archives, Record Unit 60, box 1, folder 6: no. 103, Smithsonian Institution, Meteorological Project, Records.

process of constructing their object of study. From their perspective, the conceptual boundaries of weather and what counted as information or data could be defined through exchange between volunteers and Smithsonian Institution officials.

The institution sought to define weather and data through traditional methods of scientific disciplining—such as the distribution of instructions, predesigned forms, and registers, as had happened earlier in the century as well through the proliferation of meteorological circulars. In doing so, the institution often emphasized sensorial relationships to environmental data. Observers also solidified their sensorial relationships to data in the nature of their responses. In 1852, the Smithsonian sent a circular to volunteer observers requesting "information" on earthquakes. In a letter dated May 24, 1852, which enclosed meteorological reports for the year 1851, an observer from Albany, New York, wrote: "Dear Sir, I have last night received a circular from the S. Institution asking information as to the earthquake of April last. I have replied to this in a letter. Unfortunately, it totally escaped my notice nor can I here ascertain any reliable account concerning it."[34] Importantly, the observer's language about the earthquake is oriented around the capacities of "recall"—the earthquake "escaped" the observer's "notice." In other words, the earthquake was not "felt" by the observer and therefore he could not form a "reliable" account. A reliable account would be precisely an account felt and understood by the white male body.

An observer from Frederick, Maryland, responded in a similar fashion. A letter from early June reads, "Dear Sir, The circular, dated May 5, 1852, with reference to the earthquake of April 29, 1852, came daily to hand. As the said earthquake had not been felt in Frederick, either by myself or anyone to be relied upon, I am not able to give a reply to different queries propounded. But hearing that it had been felt by my friend Mr. French, a very intelligent teacher, residing in the village of Jefferson about 7 miles SW of Frederick, I forwarded a copy of the questions. Mr. F in reply sent the enclosed answers." This observer utilizes similar language—the language of sensation and feeling—to talk about the earthquake of 1852. Because the earthquake had not been "felt" by him, or "anyone to be relied upon," he could not furnish a response to the Smithsonian's queries. Yet, word of the earthquake traveled fast, and Mr. French—a "very intelligent teacher"—in the nearby town of Jefferson provided a response to the institution's questions.

The language of Joseph Henry's "Circular Relative to Earthquakes" emphasizes the power of the body as well. The circular explains that the Smithsonian is "desirous of collecting information in reference to all phenomena having a bearing on the physical geography of this continent," and "it is respectfully requested that you furnish us with any information that you may possess, or be able to obtain, in regard to the earthquake which lately occurred in your neighborhood."[35] Henry then includes nine questions in the circular, four of which directly inquire into the sensorial memory of the observer, including, "What was the agitation felt by yourself, or by any other person in your vicinity? . . . Was there any noise heard? And if so, what was its character? Was the place of observation on soft ground or on a hard foundation near the underlying rocks of the district?" Henry's final note in the circular reads, "The direction of the impulse may have been ascertained by observing the direction in which molasses, or any viscid liquid, was thrown up against the side of a bowl." Here, Henry encourages utilizing molasses as a visual record for earthquake observation, as the "remains of the liquid on the side of the vessel would indicate the direction sometime after the shock occurred."[36] It is not surprising, then, that observers would use their bodies to help form "reliable" accounts of the weather given that the institution encouraged them to do so.

As the news from Mr. French exemplifies, requests for weather information spread through casual and informal networks of correspondence. These moments of exchange were built around shared bodily experiences of weather (here, the turbulence of an earthquake, but later, the illumination of an aurora borealis). Weather not only was a uniquely flexible category in the mid-nineteenth century but also permeated everyday social life in such a way that asserted the importance of the senses in processing the world: weather was an everyday phenomenon, felt and processed through the body. Mr. French, the reliable and "intelligent" neighbor, could become part of the weather effort simply by sharing his sensorial experiences of a weather event. In addition to deciding what exactly weather was (what materials and processes were included in the category "weather") and, more important, what it was not, observers also had a hand in deciding how weather would be studied. Just as Henry emphasized looking toward the "face" of the sky in his 1848 circular, observers emphasized the need to feel weather through the body in order to report on or record it.

Letters, circulars, and language from Smithsonian Institution reports suggest that it was a key player in constructing weather as an object to be studied, through both its circulation of directions for daily observation and notes on impromptu observation through informal circulars (later circulars focused not simply on earthquakes but also on auroras borealis and meteors). These letters also show that white male observers were key actors in suggesting and redefining what precisely weather was and how it should be effectively studied. They defined the rules of observation in their letters, and many seemed to unconsciously agree that only what was directly "felt" by the body could furnish reliable information for the Smithsonian, in addition to the observation tables they submitted.

While almost all observers were lent a basic set of instruments for their weather recording and measuring, in these early years observers and the Smithsonian Institution alike seemed to rely on the instrument of the male body to build foundational archives of data. This is significant because white maleness was idealized in relation to data. White men were idealized as data producers (as they felt earthquakes, for example) and as data managers, as instrument-based practices for collection gained popularity. Data was changing rapidly, oscillating between bodily centered practices and newly emerging instrument practices. But who exactly was allowed to participate and produce meteorological knowledge in this rapidly changing data atmosphere? Only one observer self-identified as a woman, Mrs. L. Young, from Springdale, Kentucky.[37] Mrs. Young was far from the only woman who wished to participate in meteorological data collection throughout the Smithsonian Meteorological Project. She would be joined by Anna Bowen of Nebraska Territory in 1858 and at least twenty-nine additional women from across the country beginning in 1852.

More than this, observers tended to be white and male, and some observers were active slaveholders in the antebellum region of the United States. In 1852, Maryland counted four volunteer observers—Professor W. F. Hopkins of Annapolis, Dr. Lewis F. Steiner of Frederick, Dr. William Baer of Sykesville, and Josiah Jones of Walkersville—two of whom, Baer and Jones, were slaveholders.[38] In Mississippi, both observers in 1852—Geo. L. C. Davis of Natchez and Thomas Oakley of Jackson—were slaveholders. And in Minnesota Territory that same year, Robert Hopkins was acting as a weather observer while also operating as a missionary on Dakota and Lakota lands, squatting in a dwelling on St. Peter's River with his wife, his wife's sister, and three children. The different injustices of enslavement and Indigenous land dispossession were enacted on a singular timeline, one that espoused a science for some—or a science for white male settlers—at the expense of others. As the next section explores, white women both struggled to gain access to and simultaneously aided in settler science, furthering the injustices on which settler science was built.

WOMEN IN THE FIELD: BOWEN, YOUNG, AND BAER

Anna Bowen had been misled. In 1849, word of the meteorological project was circulating through regional and national channels, from newspapers to reports and circulars. In March 1849, the *Semi-Weekly Eagle* from Brattleboro, Vermont, reported that "the Smithsonian Institute will assume the burden of furnishing the necessary instruments to those, who are unable to do it themselves."[39] The Smithsonian circular—which noted that "blank forms free of expense will be sent to those who are disposed to join in the observations, & as soon as the amount of funds for this purpose is sufficient, full sets of instruments will be furnished to careful observers in important localities"—had been "distributed to the several parts of the Union through members of Congress" starting as early as 1848.[40] By 1852, the Smithsonian announced in its annual report that a "set" of instruments "consist[ing] of a barometer, thermometer, hygrometer, wind-vane, and snow and rain gauge" was available for purchase at a "reasonable price," or on loan, from the institution.[41] By 1856— just three years before Bowen would write to the project to request her rain gauge—the Smithsonian announced that "an appropriation has also been made for the purchase of a large number of rain-gauges, to be distributed to different parts of the country for the purposes of ascertaining more definitively with compared instruments the actual amount of rain that falls in the different sections of our extended domain."[42] What did distribution look like in practice? Who would, according to the institution, be counted as a "careful observer"?

In this section, I investigate women's requests for and access to scientific materials—such as meteorological forms and rain gauges—arguing that, unlike their male counterparts, women often waited longer for tools and instruments when they requested them from the institution. This difference signified who could participate in popular science, including weather data collection and calculation. I also survey the variety of concerns that women of the Smithsonian Meteorological Project expressed through their letters, highlighting several of the main characters in the life span of the project. As meteorological practice moved from state-based systems to a national one, men were favored as data collectors and calculators despite the fact that the national system was supposedly open to all "friends of science" regardless of gender.

Before considering Bowen, and several other women like her, more closely, it is important to understand what the Smithsonian Meteorological Project was and how women created and participated in a distinct type of scientific culture related to gender norms at the time. The project was established in 1848 by Joseph Henry, the first secretary of the Smithsonian Institution, and ran until roughly 1870. The project gathered data— temperature, barometer, rain, cloud, and severe weather data—from volunteer observers across the nation. From 1849 to 1859, twenty-nine women volunteers from across the eastern and central United States collected data for the project and reported to the Smithsonian under female names.[43] The geographic distribution of these observers extended no farther west than Nebraska Territory and closely mimicked settler patterns.

From 1860 to 1870, forty-six additional women joined the Smithsonian Meteorological Project as volunteer observers, roughly 25 percent more than in the previous period, with two observers reporting from as far west as Spanish Ranche and Honcut, California. In this period, we find a large number of women reporting instrument readings to the institution—Julia Child, Charlotte Rockwell, Miss Bella Moore, Lillie Thrift, and Miss Horn observed the weather with a thermometer and barometer; Anna Spencer, Rebecca Sheppard, Miss Proctor, and Mrs. Ryerson took to the field with a thermometer, barometer, and rain gauge.[44] By classifying women's labor practices, affective concerns, and forms of meteorological data production, we see how white women engaged the Smithsonian's open call for volunteers, sending multiple requests for materials and information when their first, second, and sometimes third requests went unanswered. Despite the fact that more women were reporting with instruments in the second period of the project, studying individual letters and experiences also reveals the many hurdles that women encountered when trying to gain access to the communication networks and tools needed to fully participate in the project.

When Susannah Spencer wrote to the Smithsonian in 1853, a year after she had begun observing for the institution, she offered her records of "periodical phenomena of animal and vegetable life" freely.[45] Despite this fact and although Spencer had volunteered her services in 1852 and 1853, she was never included in the official list of volunteer observers. While some women received credit for their work, others did not. Maria Mitchell, the nineteenth-century astronomer and then librarian from Nantucket, Massachusetts, would have a similarly frustrating experience with the institution. Mitchell, who had been keeping her own meteorological records at the Nantucket Athenaeum, received blanks from Joseph Henry on September 10, 1855. Mitchell's name does not make the official record, although she clearly was involved in meteorological observation. Instead, her husband's name is listed in the observer record for Nantucket, 1853–1861.[46]

Women's letters were often marked by fear that their observations and calculations would be full of error and therefore not useful to the institution.[47] An 1856 letter from Mary Goff read: "Enclosed are the tables for the last two months. . . . Please do let me know if there are any alterations to be made in the tables." From Goff's reference to the possible need for "alterations" to women observers alluding to their "mistakes," women were cautious participants in the meteorological project, fearful of the quality of their work. This observation is valuable because it speaks to how women constructed their identities as workers rooted in their own perceptions of the values of their labors. Though more women were observing the weather with fuller sets of instruments, this feeling remained a vital part of what it meant to be a woman observer through the late 1850s and 1860s. In 1859, Sara Thomas wrote, "I am aware that any reports are open to many criticisms and have feared that you might find them of little or no use. Should they be of sufficient value for you to wish for their continuance, I shall be obliged to ask for another supply of blanks."[48] Despite the fact that throughout the meteorological project period women were active contributors to the meteorological data landscape, they often feared that they would produce useless or bad data.

While women were undoubtedly capable of producing meteorological materials that were just as valuable as that produced by their male counterparts—and sometimes more valuable, as Mitchell's life's work shows—they faced significant hurdles brought about by a scientific culture that marked male bodies as the idealized body. This sexism, which was a general attitude of the time, inflected women's own understandings of their observations, as they often worried they had produced something that would be incorrect or not useful in the institution's eyes. Even more, as the case studies that inform the remainder of this section suggest, the veracity of data was often connected to gendered bodies: female bodies were not historically marked as the arbiters of scientific data.

When Bowen wrote to the institution to ask for a rain gauge, her request was not out of the ordinary for what the institution might expect. In fact, the institution had been fulfilling instrument requests throughout the 1850s, keeping detailed records of instruments lent to volunteers. In "Meteorological Instruments Loaned, 1850–1870," we see that Dr. Reverend Hamilton from Bellevie, Nebraska, received one Kendell thermometer and one five-inch rain gauge in 1857. Dr. John Evans was provided with ten thermometers. Though Evans was an unreliable primary observer throughout the late 1850s, he was tasked with "distribut[ing]" instruments "on the route to Oregon" in the early part of the decade.[49] Bowen's request for a rain gauge not only had to be read and understood as legitimate in Washington, DC, but it also had to be understood as legitimate by the likely-to-be male distributor of the instruments within the state. Though she was clearly interested in the rain—as she drew rain lines in an attempt to measure rain, as a gauge would, on her blank form in 1858—Anna Bowen's requests for a rain gauge went unfulfilled, and she was therefore not able to cultivate her scientific interests as could male observers with access to instruments at the time.[50]

Institution records show that Bowen was seventeen years old when she began observing. She continued to report to the project up until 1864 under the name Miss Anna M. J. Bowen, at which time she married John Smiley. Bowen stopped recording and passed along weather records to her brother John S. Bowen. John utilized a thermometer, the same instrument that Anna Bowen was equipped with over the previous eight years, to perform his meteorological duties. His requests for blanks were met efficiently as he continues to write to the institution throughout the 1860s. On November 30, 1866, J. S. Bowen wrote, "Your observer was compelled to use the whole of the blanks last sent to make of his weather for the ensuing year. There were no envelopes sent with them. Oblige by sending supply of both."[51] A clerical note on the letter reads, "Sent, December 28th."

White women were deeply connected to and dependent on the men in their lives (primarily husbands and fathers) while they were pursuing meteorological science and volunteer work. At the same time, volunteer science was on the rise in the United States and was especially common as a familial activity. For this reason, it is challenging to discern who wrote for whom, who collected observations, and who simply signed letters. The line between assistant, aide, and collaborator could be blurred. This possibility of blur in scientific roles within the sphere of meteorological volunteerism is what allowed women to enter the community. As early as 1849, at least one other woman had reported as an observer to the project— Mrs. L. Young. Mrs. Young was noted as the only woman observer for the year of 1849. She was the sole observer listed in the Smithsonian meteorological records, yet all her letters were signed "L. Young," utilized male pronouns, and seemed to be written from the perspective of Mr. Lawrence Young, a wealthy farmer and slaveholder.[52] Did Mr. Young write the letters while Mrs. Young conducted the observations? Or, was Mrs. Young simply the notetaker while Mr. Young conducted observations and wrote the letters? It is possible, one might speculate, that Mrs. Young was in charge of the enterprise the whole time, not only performing and noting the observations but also writing in the voice of her husband and signing her letters L. Young in order to be involved with the scientific enterprise. However, my interest here is not necessarily settling the issue of who really penned these letters but rather in illuminating the historically porous and slippery boundaries between data laborer roles in this period.

The Youngs' first letter to the Smithsonian emphasized interest and willingness to serve the institution in a volunteer capacity, noting that they had heard about its plan to "organize for the collection of meteorological facts," having seen the meteorological circular in a journal. They continued: "Fearing you may be deficient in collaborators, I have taken the liberty to address you alone merely to say I wish the enterprise success—and that if no better opportunity presents itself for your making a station in or near Louisville than my offer—you are at liberty to avail yourself of our observations."[53] The letter moves between the use of "I" and "our" and, while the voice in the letter offers to enter the volunteer project, the deferential turn of phrase "if no better opportunity presents itself" is used. Reports for the year of 1849 list Mrs. Young as the observer in Springdale, Kentucky, and it is Mrs. Young who remains the listed observer throughout the project period.[54] As letters show, "Mrs. Young" needed materials for observation throughout the 1850s. While it is not clear whether Mrs. Young was writing on behalf of Mr. Young, or vice versa, it is clear that Mrs. Young was involved in meteorological measurement and that the nature of her involvement was intimately tied to her marriage to Mr. Young.

Much like Bowen, L. Young and Mrs. Young struggled to receive materials from the Smithsonian. In 1850, L. Young requested information about procuring a barometer for the second time: "Some time since I wrote you a few lines inquiring from what source I would be most likely to obtain an accurate barometer—Since that writing I have been induced to think J. H. Temple of Boston a skillful manufacturer."[55] L. Young went on to request advice on this matter, should the institution have any. L. Young also notes in their letter that Mrs. Young has run out of blanks. On her returns for July or August 1851, Mrs. Young noted that she needed a new supply of forms.[56] In September of that same year, L. Young wrote two letters to the institution: on September 2, L. wrote, "Mrs. Young is without blanks for additional numbers of her series of Reports."[57] And again on September 30, "In forwarding the last monthly reports—Mrs. Young stated the fact that her blanks were exhausted—two months remain now unreported for want of blanks of which I take this opportunity to appraise you." On January 10, 1852, L. Young wrote, "Gentlemen, Having heretofore signified by two letters at different periods that the blank forms for meteorological reports were exhausted and asked for renewals if copies of our tables were desired I was led to conclude you had met with some neighboring reporter more to your acceptance and had dispensed with ours."[58] The institution did finally respond to L. Young, sending blank forms and information about a James Green barometer, but it waited a while to answer—was this because the primary observer was a woman? Is it possible that Mrs. Young's blank requests were taken less seriously because they arrived without L. Young's letters or, put another way, without a male sponsor? Was Mrs. Young the Anna Bowen of the early 1850s?

Mrs. L. Young, one of the longest-serving observers within the Meteorological Project, volunteered her efforts from 1849 to 1868.[59] While Mrs. Young's name was replaced by Mr. L. Young from 1851 to 1855, her name returned to the books in 1856. In the Smithsonian Institution's 1873 annual report, which included a comprehensive list of all previous volunteer observers, Mrs. Lawrence Young was noted as the sole observer for Springdale, Kentucky, from 1849 to 1868.[60]

Entering the project earlier than 1854 might have proved challenging for women given the views of Lorin Blodget, who was hired by Joseph Henry in 1851 to reduce meteorological data and prepare meteorological reports for publication. Blodget was not only responsible for reducing observer data but also was in charge of "conduct[ing] the institution's meteorological correspondence with the volunteer observers" from 1851 until he was dismissed from the project in 1854 over a dispute about meteorological data ownership.[61] This meant that any women observers wishing to be part of the meteorological project would have their letters received by Blodget, who had been frank about the people he believed belonged in Henry and Espy's "friends of science." His public remarks on scientific matters included the following: "Only the best minds can successfully undertake scientific calculations and computations; and these must possess a half masculine strength and endurance."[62]

There were clear limits to who exactly could participate in volunteer observation and data production, and that must influence how we read the historical participation of women in this period. In theory, the Meteorological Project should have been open to any person wishing to act as "a friend of science." But in practice, the project was predominantly white and male, and its borders were officially and unofficially policed by men enforcing social norms undergirded by overt and non-overt sexism (figure 2.4). Men in power were creating new norms around gendered scientific labor that preserved their dominance, and women sought to gain admission to the work of volunteer science through male networks.

Blodget was one institution official responsible for answering the Youngs' letters between 1851 and 1854. It was no coincidence that Mrs.

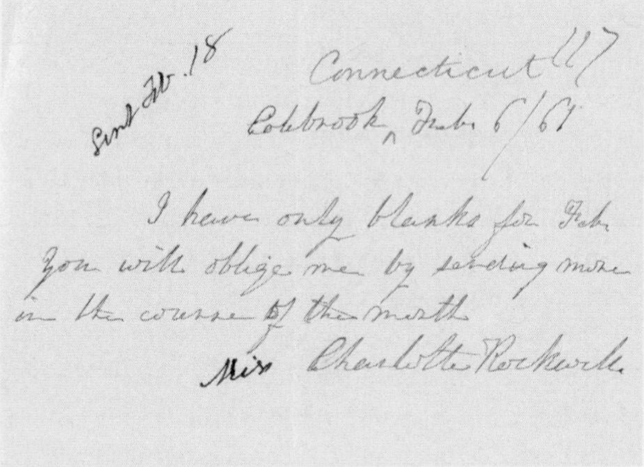

Sent Feb. 18

Connecticut 117
Colebrook, Feb. 6/61

I have only blanks for Feb.
You will oblige me by sending more
in the course of the month
Miss Charlotte Rockwell

FIGURE 2.4. Letter from Charlotte Rockwell to Smithsonian Institution officials, who added a "Miss" to her request for blanks in June 1861. Courtesy of the Smithsonian Institution Archives, Record Unit 60, image SIA2022-000497. Smithsonian Institution, Meteorological Project, Records.

L. Young was blotted from the record when Blodget began handling correspondence for the project, and that she returned to the record book in 1857, three years after Blodget's firing. Henry's views on women were more progressive than Blodget's. Henry was involved in preparing a report that reviewed the Rutgers Female Institute in 1842 and assessed the role of the sciences in female education. He conducted an "examination" and concluded that "the Committee have been strongly impressed with the importance of the Mathematics in female education."[63]

Much like Mrs. Young, Harriet Baer of Maryland would utilize male aid in becoming and remaining a part of the meteorological network throughout the 1850s and 1860s. The Baers were slaveholders: although they measured the weather and sought to provide scientific experience for their white daughter, they also upheld the system of enslavement.[64] Harriet Baer was born around 1820 in Maryland and began writing as a volunteer observer from Frederick, Maryland, in 1851. Much like Young's, Harriet Baer's labor was uniquely attached to a male figure, her father, William Baer, an agricultural chemist. While letters from the period reveal that both William and Harriet Baer were active letter writers to the Smithsonian Institution, Harriet often penned letters for her father, signing as "Harriet Baer on behalf of William Baer" and also ventured into solo writing herself. Harriet Baer's letters reveal again the ways in which the patriarchy—here a father's power—might act as a conduit for emerging women in science.

Although Harriet Baer had been reporting with her father since 1851, it was not until 1855, a year after Blodget's firing, that she was acknowledged as an observer in the volunteer record. She was a frequent correspondent—penning thirteen letters to Joseph Henry, ten of which can be classified as letters of request. In these, Baer requested blank forms, seeds, and clarification on delays of materials. Like Young, Harriet Baer is associated with a male member of her household as she becomes and remains an active member of the meteorological project. But more than this, Baer's letters also reveal that even when writing with the aid of the patriarchy, women observers still failed to receive basic materials from the institution and often spent much energy within their letters begging for materials. Between 1858 and 1869, Harriet Baer requested blank forms six separate times.[65] On February 8, 1858, she wrote, "I am entirely out of blanks!" In September and October 1858, she requested blanks again. On March 24, she wrote to the institution, "The fault is somewhere, but certainly not with me. I have not failed to send the records since I have been keeping the tables." On July 1, 1858, and January 1, 1859, Baer pleaded, "I'm out of blanks" and "Please send blanks."[66]

Why did Baer have to continuously beg for materials, month after month? As compared with Bowen's, Baer's labors show that working with a male member of a household afforded some special privileges in access to instruments—William Baer seemed to have been able to procure a full set of instruments either from the institution or by mail order—but the fundamental relationship of needing to beg, almost constantly, for materials when one wrote as an independent woman rarely afforded more power. Women like Baer and Bowen, as well as Young, waited longer for materials or were not sent materials at all, though their male counterparts at the time were. Other male observers in 1858 and 1859 were receiving blanks without issue. Because they were not needing to write letters begging for materials, they were able to focus on scientific matters rather than on exasperated attempts to request paper and instruments.[67] While it is true that this was a transformational and overwhelming time for the Smithsonian, and that male observers were also waiting for materials, women were often waiting longer, and at times their requests went unanswered. J. S. Pashley of Osceola, for example, wrote on June 1, 1860, "I wrote for more envelopes and two—Herewith goes the last of those and now I require both blanks and envelopes." A note on the left-hand side of the letter reads, "Send blanks and envelopes; Sent."[68]

The internal rhythms of the Smithsonian were ripe with processing delays, data scandals, and struggles for funding. Henry was assisted in the early years of the project by Edward Foreman and Lorin Blodget. Though the two men were assigned different tasks, their roles overlapped. As Henry's general assistant, Edward Foreman "handled routine meteorological correspondence" until he left in 1853.[69] Lorin Blodget was hired in 1851 to "reduce and discuss the meteorological observations collected by the Smithsonian Institution."[70] After Blodget was dismissed from the project in 1854 for misusing observer data in his own research, Henry was overseeing the project by himself.[71] When Joseph Henry pleaded for public patience in his 1856 secretarial report, it was because there was simply too much data to process, reduce, and send to print. Although the Patent Office underwrote the costs of reducing data, providing instruments, and sending and receiving blank forms, in 1860, support from the Patent Office was suspended, and Henry was left to manage "the whole establishment" on his own.[72]

The cases of Bowen, Young, and Baer show that women's correspondence with the institution, as well as their journeys into meteorological observation, were influenced by the men in their lives. In waiting for blanks and instruments, some women, like Mrs. Young, seem to have either employed the help of their husbands or have written under auspices of their husbands to secure entry and materials. Although enlisting the help of male power did not always translate into immediate action— after all, L. Young wrote to the Smithsonian on two separate occasions in 1851 to complain about Mrs. Young's missing material—it was one way that white women at the time used the patriarchy to gain access to a scientific world into which they were often forbidden entry.[73] The lesson here is that even when women utilized familial male power, they still waited longer than their male counterparts for meteorological materials, and this waiting likely affected their labors within the Meteorological Project generally and their capacity to shape meteorological thought in particular. Anna Bowen never contributed a rain gauge record from Elkhorn City, Nebraska. Her unrequited request translated into a nonexistent record.

Understanding how informal and formal gendered practices informed meteorological data collection means diving into how the Meteorological Project was publicized in the late 1840s, the nature of responses received in the early 1850s, and how these responses were dealt with by the Smithsonian throughout the 1850s and 1860s. Doing so reveals that women volunteers were just as eager as their male counterparts to collect meteorological data but that their practices were often stalled by the bureaucratic details of collection—procuring instruments and weather blanks. Many women observers persisted over periods of two to four years, requesting the same materials over and over again. Sometimes they received them, and sometimes they did not.

Gender-based scientific labor frameworks shaped how women were integrated into a meteorological community that had been, in the early part of the century, entirely male.[74] Uncovering lost labors reveals how meteorological data has been marked by the legacies of gender exclusivity. While women in the field were gathering meteorological data and sending it to the Smithsonian Institution monthly, another group of women were tasked with calculating the very meteorological data that women in the field were gathering. As the following section shows, women faced similar challenges in working within and for a dominantly male meteorological data culture—these women were often paid less than their male counterparts, and they were confined to hidden basements and domestic spaces to perform their calculation work.

THE COMPUTERS OF EASTON AND THE CLERKS
OF THE BASEMENT PATENT OFFICE

She had decided to take the data with her, out of Easton, Pennsylvania, and away from Lafayette College, where she had been performing calculations for about a year. It was 1857, and though her data was supposed to be bound for the newly established Meteorological Project at the Smithsonian, she had kept it for herself—seven months of numerical weather observation sheets from Swiss doctor Louis Berlandier's travels throughout Matamoras, Mexico. Yes, she had taken them and told no one. And she would keep them with her, throughout at least one move, for two full years. On the evening of February 28, 1859, she would return the data sheets to James Coffin at Lafayette College, who would then write to Joseph Henry at the Smithsonian and inform him that the missing data had been returned.[75]

Maybe her name was Maria Kutz, the only census-registered clerk in Northampton County in 1860. But she also could have been Susan R. Miller, Miss E. H. Hoagland, or Miss M. M. Shattuck, all assistant teachers in Easton public schools in 1857 who were likely turning meteorological observations into national climate data for Joseph Henry's meteorological project.[76] Though we do not know her name, we do know that the tables had been given to her by James Coffin, who had tasked her with aiding in the reduction of twenty-five years of observations—temperature and rain readings, cloud coverage and wind direction. The unnamed woman was in possession of the weather data because she had been tasked with "preparing" it for Coffin, Henry, and the American public.[77] And she was not alone in this task. From roughly 1856 to 1870, twelve to fifteen women from Easton, Pennsylvania, were tasked with the work of meteorological data reduction.[78] Called "assistants" as well as "computers" throughout correspondence and official Smithsonian publications, their work assumed skill in both copying and computation (copying observations from weather tables into a standard table form as well as calculating daily averages from Coffin's thermometer and barometer readings).[79] The work of these unnamed women would help Coffin contribute sets of data abstracts and summaries to the Smithsonian, some of which were deposited under "materials . . . relative to the climate of the United States."[80]

Women became an integral part of the metrological project between 1847 and 1870 not only as data collectors in the field but also as human computers. Much like the women in the field, the Easton computers were able to practice data collection from their homes, a tactic used by many nineteenth-century women who were forbidden from sharing physical spaces of scientific labor with men, including Clara Barton and Maria Mitchell—two women considered here—as well as Elizabeth Lomax.[81] The computers of Easton were hired through James Coffin—though Joseph Henry was aware of the arrangement—who had been utilizing the labors of his daughter, since at least 1840, as a copyist for *Winds of the Northern Hemisphere.*[82] The reason women were invited to enter meteorological computing was due to public frustration around the lack of weather data that had been published and made available to the public by the Smithsonian Institution. In his 1856 secretarial report, Henry pleaded for public patience: "Complaints have been made that but few of the materials collected by the Institution have yet to be published . . . it is more important that the information should be reliable than that it should be quickly published."[83] Henry's remarks suggest a situation of environmental data overload. The fundamental solution to that overload—and the public's impatience in receiving data—was to employ women cheaply, thus allowing them to enter meteorological data reduction only through settler masculinity norms.[84]

In early months of 1855, Henry and Coffin addressed the issue of data overload. Henry wrote to Coffin that the problem could be the employment of "students" (seemingly genderless, but all students would be college-aged men at the time) as assistants. On March 16, 1855, Coffin wrote to Henry that he had "accidentally found a very efficient aid" to help with the "reductions of current observations in the form of the Superintendent of the Public Schools of our Borough."[85] Coffin had connected with the superintendent in an effort to see whether any of the teachers were "careful and accurate computers for such kind of work" when the superintendent asserted that he could take on such work himself.[86] Though Coffin recognized that "the position he now holds is a responsible one," Coffin expressed "surprise at his offering his services so low" and continued that "some of the simple operations such as transcribing the results into tabular form I suppose might be performed just as well by females."[87] In July, Coffin wrote again to Henry, noting that "the general reductions are progressing. I have over a dozen persons employed upon the work, mostly teachers of schools in the borough, who have now a vacation of two months."[88] Though Coffin does not specify gender in his letter to Henry in 1855, Henry announced in the Smithsonian Institution report of 1856 that "twelve to fifteen persons, many of them females, have been almost constantly employed, under the direction of Professor Coffin in bringing up the arrears and in reducing the current observations."[89] By 1857, women were employed in more than the "simple operations"—they were involved in correction and reduction.[90]

In addition to reducing Louis Berlandier's observations, the women of Easton, Pennsylvania, joined male computers in the work of reducing one of the first publicly available climate data resources in the United States. They were responsible for bringing their calculation skills to bear on the Smithsonian volunteer tables, the very tables generated by observers like Bowen, Young, and Baer.[91] In 1856, the computers of the Smithsonian Meteorological Project were responsible for reducing "upwards of half a million of separate observations," "allow[ing] an average of one minute for the examination and reduction of each observation."[92] The practice of reduction included, first, an examination of each "observation" contained in the table and, second, "arithmetical calculation." Each observation form asked observers to record barometer, thermometer, and hygrometer readings, as well as to record rain and snow measurements, cloud coverage and direction, wind direction, air pressure, and humidity. Each observer was required to make three readings per day—9:00 a.m., 2:00 p.m., and 9:00 p.m. So, a human computer would sit down with a pile of these forms and perform "arithmetical calculation" in order to yield the following results: the "mean, maxima, and minima" of a month's worth of observations. The process of reduction was a repetitive venture of addition, subtraction, and averaging. As Henry outlined in his 1857 report, the computers would have to perform these reductions over and over again on half a million observations in a given year. But they were not at work on just one year of collected observations. They were working through five years of reductions, which in the end, would likely total over two million observations in need of reduction.[93]

Over the seven-year period that Coffin headed the effort to prepare Smithsonian meteorological observations for publication, he utilized the labors of male and female workers.[94] In hiring women computers—drawn not from male-only Lafayette College but rather from the Easton public school system—Coffin emphasized the need to "hire them as cheaply as possible."[95] There was too much data to process and not enough labor or money to pay for the processing. In securing a partnership with the Patent Office, Henry had made several things possible. First, he secured funds from the Patent Office for the project so that workers—including Coffin—could be compensated. While Coffin and Henry agreed to pay workers twenty-five cents an hour, there is no evidence that male and female computers were paid the same wages for their work.[96] Coffin did not submit a list of his employees to Henry. This partnership also meant that the Patent Office was doing more printing and mailing of materials on behalf of the Smithsonian, including printing blank forms for observers and mailing them under the office's franking privilege.[97] The same year that Henry and Charles Mason (commissioner of the Patent Office) began their partnership—and during which Mason offered printing services to the Smithsonian—Mason hired his first cohort of full-time women clerks, including Clara Barton, at "ten cents a hundred words," which was the going rate for male clerks at the time.[98]

Between 1854 and 1865, Clara Barton, who "had perfected a 'copper plate' style of handwriting," was legally employed as a clerk at the US Patent Office.[99] Barton was invited by Charles Mason to operate as a clerk "copying patent applications, caveats, and regulations."[100] Her clerk work included copy work from the "Department of Agriculture, Smithsonian Institution, and the Weather Bureau."[101] In addition to being employed as the first official female clerk in the Patent Office, Clara Barton is important to this story because she represents the final node in a network of women laborers who strove—and sometimes failed—to gain access to scientific and meteorological labor in the period. The network extends from women collecting meteorological data across the nation, to women assistants reducing data under the guidance of James Coffin at Lafayette College, to Barton's office of ladies copying these data sheets for a congressional publication at the Patent Office. Different nodes within this network were given different levels of flexibility to perform their duties as data gathers and compilers.

From Bowen to Barton, it was precisely the ways that scientific labor was connected to the body that produced gendered data collection and computation practices. Though historians of American meteorology have tended to think of the Smithsonian Meteorological Project as a democratic science project, Bowen and Barton show that the project evolved under a gender-typical regime of labor that privileged male scientific labor. Clara Barton and her cohort of women clerks were "confined to the basement of the east wing of the Patent Office," where they did not have to "come in contact with the public."[102] Although the ladies had been confined to the basement, their physical presence in the Patent Office building was not welcomed by all. Mason left the Patent Office temporarily on July 5, 1855, for four months. On July 9, then acting commissioner of the Patent Office, S. T. Shugert, moved Barton and the other full-time female clerks to temporary status.[103] In addition, Shugert wrote in a letter to the chief clerk of the Interior Department: "I have communicated to the Ladies employed in the Patent Office that they must vacate their room within the present month, and also, that the Hon. Secretary promises to give them work from the Land Office, when practicable."[104] Women clerks were asked to vacate their basement offices—they were being phased out. Robert McClelland, secretary of the interior, agreed: "I have no objection to the employment of the females by the Patent Office, or any other of the Bureaus of the Department, in the performance of such duties as they are competent to discharge, but there is such obvious impropriety in the mixing of the sexes within the walls of a public office."[105] The problem was not a woman's capacity to do the work assigned—or so they said— but rather that the physical presence of women threatened the fully masculinized practice of male work. On behalf of preserving scientific and social space where "only men could be comfortable," it was clear that the women had to go.[106]

From Easton to Washington, DC, there were valuable differences in male and female data production. Namely, because men were allowed to publicly claim their connections to data practice and production, they were also allowed to work *publicly* on data reduction. Male computers were able to work on their reductions at Lafayette College in the Star Barn—the meteorological building on campus—or in the Mathematics Building, the Patent Office, or the Smithsonian Institution offices. Women were often required to perform their reductions at home. This was the case not only for the Easton ladies but also for Clara Barton and at least three additional women "clerks" who set the data reductions from the Easton ladies to print for Congress. While Clara Barton and her crew began their work in the basement of the Patent Office (consigned there so that they did not come into contact with male Patent Office workers), they were forced at times to do their meteorological copying work at their homes. In the late summer of 1855, after Shugert and McClelland voiced their disdain for "mixing of the sexes within the walls of public office," Barton "clean[ed] out her desk and work[ed] from home as a copyist."[107] When she returned to her desk in the fall of 1855, after Charles Mason returned, "a campaign of harassment by some of her coworkers evidently began." All her coworkers were male.[108]

THE LABOR OF DATA MAKING

The weather data that women in the field, in Easton, and in the Patent Office basement helped create was printed and used, according to Henry, for isothermal and rain mapping of the country. Their work led to the first published account of weather and climate data in the United States based solely on national volunteer data.[109] Taken together, women meteorological computers in Easton, clerks like Barton, and observers like Anna Bowen are constellatory stars in the scientific labor system that defined mid-nineteenth-century scientific work for white women. Locating this constellation recenters nineteenth-century meteorological and science culture away from exclusively male narratives and toward the underexplored roads that women took to shape meteorological science, weather archives, and public data at the time. Barton's story of struggling to maintain her appointment as a clerk while remaining visible within the male workforce in Washington is illustrative of a larger dynamic at play in this period that has to do with the expression and regulation of data (whether in the field or in the office) through the politics of the body. Barton's expulsion from the Patent Office was due to the way her body had been marked by gender, despite her incredible skill; Coffin's employment of "cheap" female labor was founded on the practice of paying gendered bodies less; Blodget, Shugert, and McClelland saw the female body as a threat to male spaces of labor and science, thereby policing them out of practice.

Although one's gender did not make for the production of a different type of data, weather data collection and calculation practices in the mid-nineteenth century were tied up in power structures that made it impossible for women to practice environmental data collection and reduction as men could and did at the time. While women "entered civil society in the 1790s and in increasingly large numbers in later decades," as historians of women's labor note, women were not able to practice data collection and calculation in the same ways that men did at the time.[110] Women struggled to claim their visibility within the public sphere because of what their bodies symbolized or advocated.[111] This is what we see in the meteorological data atmosphere: women were doing the work of making weather data, reducing data tables, and copying data blanks without being able to claim their practices publicly. When they did attempt to do so—in letters to the Smithsonian and in Patent Office basements—they were seen as threatening established systems of male dominance. These threats reveal that the culture of weather data production—from the field to Easton and the basement of the Patent Office—depended on a gendered labor system.

This chapter reveals the ways in which white women both struggled to gain access to and simultaneously aided in settler science, often participating in, and advancing, settler colonial logics—including Indigenous land theft. This participation and advancement allowed for the material practice of settler science. Although the data labor of white women broke the boundaries of white male science, white women are also complicated—compromised and complicit—figures within the longer history of US settler science, colonialism, and the institution of slavery. Additionally, in order to enumerate this history of women in science within a larger period bound to male dominance, it was necessary to work with murky records and blurred accounts. Unfortunately, this was—and is—precisely the point. Yet, even these women left a trail, as overgrown as it may be. I hope that my efforts have cleared the overgrowth a bit so that someone else might find their way down what is likely a much longer path.

3

DATA IN THE SKY:
SCIENTIFIC KITES,
SETTLER MASCULINITY,
AND QUANTIFYING THE AIR

"This month being considered as one of the best for flying kites, we may indulge our young friends with an article on the subject," wrote *Scientific American* in its third monthly installment on October 10, 1846, the year of its first issue.[1] The article describes "some fancy models of kites, which are not often seen," including an eagle, a rose, and a star kite, and then moves on to instruct the reader in construction: "To construct this figure there must be four light rods of wood, made to cross each other in the centre, being there lashed together, and thus constituting eight arms. . . . The twine must be fastened to four of the arms, and the tail of the kite should be covered with green paper, which by the contrast will have a pleasing effect."[2] The article focuses on home construction and construction of kites that imitate nature, inviting "young friends" to bring nature—the bird, the rose, the star—into animation. Much like James Espy's "friends of science," the imagination of who might act as a "young friend" in this context was predominantly white and male.[3] In its inaugural set of issues alone—which spanned August to December 1845—*Scientific American* published four articles that centered white working-class masculinity, including articles such as "Thoughts for the Working Man" and "The Wild Man of the Mountains," which positioned white male science as the idealized *Scientific American* subject. The importance and idealization of the white, working-class, male scientist was defined against the backdrop of traditional ecological knowledge—which the publication describes as unscientific—and rich white elitism.[4] Popular science venues, such as *Scientific American* and the *New York Times*, aided in solidifying a masculine settler science across the middle to late nineteenth century.

The kite was a site of masculine settler science; across the century, it was an emerging technology and a boy's toy. Kites not only reproduced the natural world—and, in their construction, acted as pedagogical tools that connected imitation of nature to leisure and play—but their construction and flying were linked with a masculine history of scientific discovery.[5] In the 1840s, *Scientific American* encouraged young boys to build bird and star kites; across the same publication, they would read about Benjamin Franklin's kite experiments. Later in the century, they would utilize these kites as meteorological tools for exploring the upper air. Through the US government—the Weather Bureau and the War Department—and the work of white elites and amateurs across the country, kites helped to develop what Jeanne Haffner has referred to as an "air-mindedness," a fervor for aerial views. But where Haffner's air-mindedness was rooted in aerial views from aviation, this air-mindedness was rooted in settler masculinity and optical meteorological data capture.[6]

The militarized colonialism of the nineteenth century enlisted settlers in biospheric domination and mastery of nature, peoples, and environmental knowledge systems. Settlers learned to value environments and peoples through the activity of conquest and domination, which led to the erection of settler infrastructure on lands that did not belong to them. This included building and maintaining military-meteorological forts and telegraphic infrastructure from 1870 on in regions such as Montana, Wyoming, and Dakota Territory for the express purpose of strategic data collection and transmission.[7] Weather forts were set up across Arizona for the purposes of settlement, weather data collection, and the attempted dispossession of the Apache and Hopi peoples of their homelands.[8] Settlers produced environmental knowledges, in the form of data, about these places that were informed by values of heteropatriarchal domination, theft, enslavement, and genocide.

In this chapter, I explore how settler weather data cultures—specifically, upper-air remote data sensing and transmission technologies of the late nineteenth century—extended values of heteropatriarchal domination and masculine science. In this way, I explore how, in the words of Jennifer Gabrys, "sensors inform our engagements with environmental processes and politics."[9] The first section of the chapter investigates the status of kites as toys and meteorological tools in settler culture across the nineteenth century, exploring how a "boyish pastime" became a popular scientific instrument that mobilized a movement to quantify the air. The second section explores a specific late-nineteenth century experimenter, William Abner Eddy, and his partnership with scientific elites, the War Department, and the Weather Bureau. This section also uncovers print culture's role in the formation of a settler-defined meteorological knowledge, showing how print culture helped to form a settler literacy that was textual, visual, white, and male. The final section considers the technologies that allowed for meteorological kite data collection and the significance of these technologies for visualizing the nation, settler notions of data, and settler environmental consciousness in general.

As the previous chapter explored, white masculinity deeply shaped the data culture of the early to mid-nineteenth century. This intensified in the later part of the century as white male experimenters, elites, and predominantly white institutions sought to expand data collection to the air and relied on national publications to popularize their efforts. In *Imagined Communities*, Benedict Anderson established the capacity of print to create "languages-of-power."[10] According to Anderson, in the "convergence of capitalism and print technology . . . a new form of imagined community" is formed," which sets the stage for nation-building. Although Trish Loughran charts a more fragmentary and contested path toward "nationalism" through the print cultures of 1770 to 1870 in her book *The Republic in Print*, both Anderson and Loughran recognize the capacity of print culture to assist in nation-building.[11] As Monika Krause has studied, the period between 1890 and 1914 "brought a rapid expansion of the newspaper industry" where "the number of newspapers increased by one-third between 1892 and 1914, the average circulation of papers doubled and the volume of advertising increased by three-and-a-half times."[12] Given this rapid expansion during the period in which scientific kites combined with masculine science and military meteorology, this chapter contends that print culture uniquely supported the formation of a nation-based settler literacy that was founded on settler-defined meteorological knowledge.

This new literacy reoriented where and how settler communities and institutions were measuring and understanding the weather, upholding a way of relating to weather—and the newly established government entity the Weather Bureau—that valued relation through quantification, settler masculinity, and optical sensibility. Photographic technology developed alongside aerial weather data collection: weather kites were mounted with cameras, thermometers, and barometers, and this combination shifted settler perspectives of weather broadly and weather data specifically. Experimenters worked with aerial photography technologies and sought to publish aerial drawings and lithographs. National science magazines and newspapers supported these ventures, building a public discourse around this form of weather data collection. Additionally, air exploration and data collection through meteorological kiting produced distinct technologies that reframed the access to and availability of weather data. Importantly, locating data in the sky promised a continuation of the settler colonial practices that had grounded much of the century. Much like the land, the sky, too, was approached by settlers as a "frontier."[13]

As the kite traveled from the hands of the "young friends" in the 1840s to those of the experimenter in the 1890s, settlers utilized it as a technology to facilitate a search for upper-air weather data. This search would alter data cultures—the material practice of data collection, collation, reduction, and dissemination—for settlers near and far as well as for the century ahead. By the end of the nineteenth century, settler territory was newly conceivable and bound by weather data: a century's worth of settler data combined with mapping techniques to visualize the nation. The settler environmental consciousness that sprouted from these changes understood the biosphere as data-centric, quantifiable, and relatable through measurement. What was countable could be quantified; what was quantifiable could be known and claimed. This data culture of the late nineteenth century was distinct from those of the earlier part of the century in several ways. First, this data explored the lower atmosphere, a wholly new space for data collection. Second, popular science publications supported this data culture (and the tools and techniques used for collecting this data) and thus supported the formation of a settler-defined aerial meteorological knowledge bound to the same central figures from early in the century, white men.[14] Third, new tools and techniques for lower-atmosphere exploration reconfigured the concept of data by marking it through automated and remote technologies that were deeply connected not only to the practice of masculine science but also, now, to settler militaries.

THE MOST ROMANTIC DISCOVERY

Kites occupied different parts of settler environmental consciousness across the nineteenth century. In the early part of the century, they were scientific tools for settler meteorological experimentation among white elites. Outside of the Philadelphia City Hospital in 1835, the newly formed Franklin Kite Club gathered. Its members would meet there weekly from 1835 to 1856, having chosen a name for their club that signaled their affinities for Benjamin's Franklin's kite experiments. Their club had been formed "for the purpose of making electrical experiments" through the use of kites. The kites were square and made of muslin or silk cloth that was wrapped around a frame of cane; they flew on a wire of copper attached to a reel, which was itself placed on glass supports. James Espy, a meteorologist, was a member of the kite club, and on July 19, 1836, he wrote to Joseph Henry, then a Princeton University professor and soon to be the Smithsonian Institution secretary: "Dear Friend, I wish you would come down and spend Thursday, Friday, and Saturday with us, and assist us in our *kite experiments*. We have now about 3 miles of wire on our reel and 30 kites ready to take it up whenever the wind favors." Henry attended a kite-flying gathering in Philadelphia and witnessed "experiments on atmospheric electricity."[15]

The Franklin Kite Club used kites to investigate atmospheric electricity, and these investigations led to curiosities about meteorological issues, specifically clouds. In his 1839 "Essay on Meteorological Observation," Joseph Nicollet wrote, "The Franklin Kite Club, at Philadelphia, have lately discovered that in those days when columnar clouds form rapidly and numerously, their kite was frequently carried upwards nearly perpendicularly by columns of ascending air; and they say, in their report, that this circumstance became so familiar during the course of their experiments, that, on the approach of a columnar cloud just forming, they could predict whether it would come near enough to affect their kite."[16] Though the Franklin Club columnar cloud report has not been found, Nicollet's reference here is enough to suggest that settler scientific thinkers—like Espy and Henry—were interested in ways of determining the meteorological conditions of the upper air even before they got a sense of the national conditions of lower-level terrestrial air through the Smithsonian Meteorological project volunteer observers.

At this same time, kites were also toys—albeit dangerous ones—for young boys. On a Saturday afternoon in March 1856, John Gilder was flying his younger brother's kite on a roof in Flushing, Queens. He fell thirty-five feet from the roof of a building to the ground: "It is supposed that in going backwards against it [the wind] to elevate the kite, he fell back over its parapet."[17] John Gilder's fate was shared by other amateur kite enthusiasts. The *Times* reported two additional fatalities in 1850 and several others throughout the later part of the century. This would include the death in 1861 of Samuel Cloberg, who fell through a skylight while flying a kite and died within hours. Edward Delauhaunt, a "young boy" from Brooklyn, New York, fell off a roof to his death in 1862, and Frank Dalton of Jersey City was run over by a wagon and "considerably injured" while flying a kite.[18] Kite deaths had spiked so high in Philadelphia that, in 1875, there was a temporary ban on kite flying in response to "a number of accidents to persons or injury to property."[19]

Across the 1880s and 1890s, kites operated as scientific instruments among settler masses. If the young boys flying kites could survive falling off their roofs—avoiding the fate of boys like John Gilder—they might grow up to be W. A. Eddy, Abbott Lawrence Rotch, or Cleveland Abbe, all key players in the settler scientific community of the late nineteenth century. These key players were young boys when *Scientific American* began publishing DIY kite articles and showcasing the histories of Franklin's scientific kite experiments.[20] Kites were captivating, dangerous, and a way of popularizing and disciplining settler masculinity at a time when settler science had been gaining popularity through institutions like the Smithsonian and across print venues such as *Scientific American* and the *New York Times*.

In the lead-up to the 1880s and 1890s, *Scientific American* positioned kites as pedagogical tools to teach settler scientific history and promote settler literacy. An 1853 article, titled "Lightning Conductors," referenced Benjamin Franklin's eighteenth-century kite experiment: "The discovery was one of the most romantic with which we are acquainted in the whole history of philosophy. How grand the noble old philosopher printer looms up before the mind's eye, standing in his sober brown coat, gazing with his calm contemplative face, upward to his tiny kite which he had raised to lure the lightning bolt from the dark thunder chariot, and lock it to the floor of mother earth. At that moment a new science was born—that of lightning conductors."[21] Another article from 1854, titled "The Telegraph" and with a similar tone, includes a reference to Franklin's kite experiment: "His experiment of drawing lightning from the heavens, by the flight of a simple kite, is the most wonderful and sublime philosophical experiment on record."[22] In both cases, the scientific kite enters the history of invention as a "romantic" and "experimental" tool. The kite figures as a "simple" and "tiny" thing that, along with the "noble old philosopher," Franklin, aided in the "most romantic" discovery, the "birth of a new science," and "the most wonderful and sublime philosophical experiment on record." Here, the kite is the "simple" tool that does the work of romanticizing settler scientific discovery.

Print culture not only situated the history of scientific invention in the United States around such settler narratives of "discovery" but also encouraged settlers to get involved in scientific experimentation on their own with similarly romantic tones. Articles about kite-flying accidents, instructions for building kites, and discussions about the kite's role in the discovery of electricity proliferated throughout the later part of the nineteenth century.[23] We also see articles from the 1880s and 1890s that outline experiments with kites specifically related to the science of the air. An 1886 article, "A Gigantic Kite," reads: "After remaining for a long time an object of amusement merely, the kite is becoming one of study for the mechanician, who finds in it a means of applying and verifying formulas related to the resistance of the air and of this contributing to the progress." Like the DIY instructions for leisure kites in the 1850s, the later article "A Gigantic Kite" is essentially an instruction manual for those interested in building and operating a scientific kite. In describing Marcel Maillot's gigantic kite, the article reads: "The kite is a regular octagon, having a superficial area of 85 square yards, and the frame of which weighs 150 pounds. . . . Two cords, maneuvered from the earth, and connected with the two extremities of the vertical line passing through the geometric center of the kite." And then an explanation of the flight: "After firmly fastening the cord, which was 820 feet in length, Mr. Maillot and his assistants lifted the top of the kite."[24] In outlining this experiment, the article provides a level of detail that would allow experimenters to construct and fly such a kite.

Kites were cheap and portable; they were intended to be used by boys and were manufactured by "girls." In popular culture, boys were able to envision a future of scientific discovery through kite play, whereas "girls" were confined to monotonous manufacturing. The 1883 article "Penny Kites" outlines processes of kite construction from a "dealer" and describes the "variety of kites" "for the boys to choose from." The article continues: "Some things made in New York are very dear—models for the Patent Office, for example, and good lamb chops, but some other things are excessively cheap, for instance, kites, which can be had for a cent a piece."[25] Like many of the other kite articles considered here, this example includes instructions for building, and it also places gendered labor— here "girls"—at the center of cheap kite construction: "The paper, cut to the right size, is piled on a table on one side of a girl. Two piles of sticks are at her other hand, and a pot of paste and a brush before her. She spreads out a piece of paper and runs the paste brush around the edge. Then two of the longer sticks are laid in the form of an X. Across the cross of the X a shorter one is laid. Then the pasted edges of the paper are folded over, enclosing the ends of the sticks." Kites not only were cheaply produced through "girl" labor but also were compact. An 1884 article, "Miscellaneous Inventions," reads: "A folding kite has been patented by Mr. Joseph Stumpp of Brooklyn, New York . . . to promote convenience in storage and transportation."[26]

The 1890s marked a new interest in the kite as an instrument of settler science, rather than a youthful element in the "romantic discovery" of electricity and a leisure activity for boys. Kites were now important tools for scientific experimentation that could inhabit, document, and measure the upper air. Published materials from the 1890s document this important shift in the use of kites in meteorological culture; they also show the extent to which the articles themselves acted as pseudo instruction manuals for amateur builders. Instructions for building and flying kites—which were seamlessly woven into scientific articles about kite experiments as well as published more directly in places like the *New York Times*—combined with public energies for the exploration of air to make the kite the perfect apparatus for settler experimenter explorations of the meteorological conditions of the air. As William Edgar Knowles Middleton writes, "The manned balloon was too expensive and sporadic a means of collecting data from the upper air, and various events contributed to the demise of this kind of scientific ballooning. In the 1890s there was a rapid development of the use of kites and captive balloons."[27] Kites offered a relatively inexpensive way to construct an upper-air apparatus that, combined with meteorological instruments such as the thermometer and barometer, would venture into unmeasured territory.

The history of ballooning in the US settler states—not attended to extensively here—dates back to the 1840s, when the War Department considered the use of balloons for aerial reconnaissance as US settlers trespassed on and occupied the homelands of the Seminole people.[28] Balloons were again considered for reconnaissance purposes during the Civil War and resurfaced again—as a meteorological tool—when the Signal Corps became responsible for the National Weather Service in the 1870s. Balloon ascents were conducted and studied throughout the 1870s under the direction of assistant to the chief signal officer Cleveland Abbe, professor Samuel King, and Signal Corps observer George C. Schaeffer Jr.[29] The overlap between military and meteorological balloons was fruitful for the Signal Service—a branch of the War Department—in this period in that the "introduction of military balloons into the Signal Corps was thus facilitated by the utility of balloons as a weather device."[30] The Signal Corps would continue to experiment with balloons for reconnaissance up until the end of the century, when the limitations of the balloon as an impactable bomb carrier with limited mobility stifled future growth.[31] As for meteorology, balloon ascents were expensive, and kites were able to reach greater heights than balloons.[32] While balloons would reappear as an important meteorological tool in the early 1900s, kites dominated the weather experimenter scene of the late nineteenth century.[33]

Kites storied settler scientific discovery and settler masculinity through narratives of "romantic" discovery, gendered play, and settler scientific adventure. The power of kites did not disappear in the 1890s, but rather developed more profoundly within two distinct communities that overlapped in valuable ways.[34] After the Smithsonian Meteorological Project, settler experimenters—shaped by their desire to further nation-building—were leading the way in colonizing the air.[35] Settler experimenters would work with settler government officials on upper-air data collection, quantifying and colonizing the sky. The data collection processes developed and visual products produced by experimenters such as W. A. Eddy were eventually adopted by government officials, becoming part of official Weather Bureau practice.

The creation of the Weather Bureau and its precise role in the machinery of government across this period are tied to the Smithsonian Institution, the Signal Service, the War Department, and the Department of Agriculture. The Civil War greatly impacted the Smithsonian meteorological network, such that in 1866 "there were only thirty observers in the eleven southern states contributing a total of 159 monthly reports."[36]

Additionally, a fire at the Smithsonian in 1865 destroyed Henry's correspondence as well as meteorological records and scientific instruments. The fire "disrupted the work of the meteorological project" as research funds were diverted to repair and fireproof the building. In 1865, Henry called on the federal government to fund a national weather service, but "congressional approval was not forthcoming." Henry would have to wait five more years for Congress to fund such efforts.[37]

On February 2, 1870, Congress passed a joint resolution that created the National Weather Service. Both before and after the passage of the resolution, Congress debated military versus civilian control of the service; by the month's end, on February 28, 1870, Colonel James Myer had been notified that the Signal Service would have the responsibility of carrying out the service.[38] The Signal Service itself was a new entity, created in 1860 to, among other things, construct telegraph lines during the Civil War.[39] But while the Signal Service knew how to construct telegraph lines, it lacked the meteorological knowledge necessary for transmitting on those lines, so Myer hired Cleveland Abbe (a former Smithsonian volunteer) as assistant to the chief signal officer in January 1871.[40] The National Weather Service remained under military control until 1890, when the Weather Bureau was established by congressional law. Meteorological services for the US settler states were transferred to the Department of Agriculture, where the Weather Bureau (formerly the National Weather Service) would be housed.[41]

Though they differed as civilian versus military branches of government, the Weather Bureau and the War Department had considerable overlaps given that both sectors had been tasked with weather data collection and forecasting from 1870 to 1900. These once "young friends" sat at the intersection of this overlap, as they worked to extend US aerial military technology and weather science. Their work deepened historic relationships between meteorology and military technology that would continue well into the twenty-first century. The culture of a boyish pastime would come to shape environmental relations for an entire community of settlers, turning the sky into territories of data to be used by the government and military officials.

On May 30, 1895, W. A. Eddy of Bayonne, New Jersey, walked up the stairs
to his attic, otherwise known as his "kite room," where he picked up one
of his many kites. On this day, he would choose a kite with a wooden
frame and then mount a camera on that frame, securing it in place. With
the midday sun rising and the New Jersey winds maintaining their eastern disposition, Eddy attached a line and a polished metal ball to the
camera shutter, walked back down his attic stairs, and then went outside with kite in hand. He walked the grassy, spring fields leading to the
waters that surrounded his home's peninsula, Newark Bay and New York
Bay. As the dregs of May were seeping into what would be a balmy June,
Eddy placed this kite-camera apparatus carefully on the ground and then
began to lay the long kite cord on the ground, too, perpendicular to the
horizontal edge of the instrument. When the weather was right, Eddy
took the end of the kite cord and began to run with it, approximately five
miles an hour. On this day in May, around 2:00 p.m., the kite slowly rose
into the air, towering like a cloud, all the while snapping photographs of
the landscape that had once contained it.[42]

William Abner Eddy, born in New York on January 28, 1850, was a
scientific experimenter who worked with government officials, from the
Weather Bureau to the War Department. Eddy's photographs marked
a moment of technological innovation—his were the US settler states'
first aerial photographs taken from the perspective of a kite with only
human assistance from the ground. These photographs were reported
by at least one newspaper as the "first-ever" aerial photographs, despite
the fact that aerial photography had been around in the US settler states
and in Europe since Nadar's famous balloon photograph experiments in
1863 in France. Similar experiments had commenced in the US settler
states before the Civil War.[43] Although balloons had also been used to
capture aerial views by camera, and while aerial photography was practiced before Eddy's experiments, previously it had not happened in quite
this way: namely, it had not been combined with settler science, such
as remote temperature and barometer data capture. In addition to this
innovation, the ways in which these photographs were read and published show an interest in quantitative, rather than purely qualitative,
readings of aerial kite photography. The photographs were understood as
scientific and quantitative material that visualized the urban landscape
with great clarity.[44]

As spring turned into summer and then summer to fall, Eddy's experiments (in both aerial photography and upper atmosphere temperature reading) gained popularity. Newspapers and science magazines began to report on Eddy's weather-kite experiments, and in July 1894 Eddy was invited to the Blue Hill Meteorological Observatory in Boston to conduct additional experiments.[45] Founded in 1885, Blue Hill Meteorological Observatory—which was built and run by MIT graduate and Boston elite Abbott Lawrence Rotch—was a valuable weather data collection space because of its location on a hill. Interest in measuring the upper atmosphere, or "observ[ing] from a location unobstructed by buildings or trees," brought these two men, and the populations they represented, together.[46]

As this section explores, settler experimenters enlisted themselves in quantifying the air and in creating wartime technology for the War Department under the rubric of weather experimentation.[47] In the late 1880s and early 1890s, settlers not yet directly associated with the War Department and the Weather Bureau began experimenting with kites affixed with weather measuring tools. Across the decade of the 1890s, settler experimenters who tinkered with aerial data collection were called on by the War Department to develop military technology alongside weather data collection. Military technology and weather data collection practices traveled hand in hand, and many "young friends" and "friends of science" were eager to participate. William Abner Eddy experimented with kite technology for the War Department throughout the 1890s and was committed to exploring how meteorological kites could be incorporated into military reconnaissance.

Lower atmospheric measurement and kite science brought together settler experimenters, patrons, and the federal government around the shared goal of quantifying the air and capturing upper-air data.[48] What had played out on the ground over the last century or more regarding the intertwined practices of settler colonial expansion and scientific capture would soon play out in the air. As a cheap tool that stood in stark contrast to the weather balloon of the early twentieth century, kites were adopted across various sectors of the nation—from white elites and white working-class communities to the War Department and eventual Weather Bureau. Additionally, their status as a gendered tool (connected to boy's play and leisure culture) made them a perfect collaborative instrument across white male communities in the late nineteenth century.

Eddy spent the early part of his life employed as an accountant, "connected with the office of the Collector of Revenue."[49] On the side, though, Eddy had been flying kites with thermometers, barometers, and cameras attached to them since 1890.[50] He had an interest in kites from an early age—he was no doubt one of the "youth" that *Scientific American* "indulge[d]" in its midcentury kite construction and flying tutorials.[51] While Eddy began as a young kite tinkerer and later adult experimenter, he would, from roughly 1891 until his death in 1909, collaborate and share his research, experiment results, and enthusiasm for upper-air measurement and imaging with emerging professional meteorologists (as exemplified by Abbott Lawrence Rotch and Smithsonian secretary Samuel P. Langley), as well as with War Department and Weather Bureau officials (including the volunteer observer turned head meteorologist for the Signal Corps and later Weather Bureau, Cleveland Abbe).

According to instrument maker S. P. Fergusson, Blue Hill Meteorological Observatory—the first upper-air observatory in the United States— flew its first kite in 1885, the year of the observatory's founding. The kites were "coated with tin foil, and served as collectors" of atmospheric electricity.[52] Similar experiments with kites were performed in 1891 and 1892 by scientist Alexander McAdie. While kites were part of the culture at Blue Hill Meteorological Observatory, it was not until Eddy's two-week visit there in 1894 that kites were employed as meteorological data collection instruments rather than as electrical conductors in the long-standing tradition of Benjamin Franklin. On August 3, 1894, an ordinary Richard thermograph was sent up on one of Eddy's self-designed Malay, tailless kites. Importantly, the thermograph was altered so that the kite might handle the weight of the instrument more efficiently: "The heavy parts were replaced by wood and aluminum, and the modified instrument, with a small basket inverted over it to serve as a screen for the bulb, weighed all together 2 pounds and 5 ounces."[53] Eddy flew five meteorological kites at Blue Hill Meteorological Observatory (a privately funded observatory near Boston) on the following day, August 4, with at least one raised to a height of 1,430 feet.

Three years before his experiments at Blue Hill Meteorological Observatory, Eddy had written to then secretary of the Smithsonian Institution, Samuel P. Langley, about his kite experiments. As it turns out, Langley and Eddy were investigating similar topics, namely, "the laws of kite dynamics."[54] On August 17, 1891, Eddy wrote: "Dear Sir, Mr. Chamute, Civil Engineer of Chicago, informs me that you are about to publish a work on the subject of plane surfaces in their relation to air pressures. I have been experimenting with hexagon kites with the purpose of reaching a great altitude." Eddy goes on to ask how he can receive a copy of Langley's work, noting that "any suggestion or discovery of yours which I may use in my own experiments I shall be very careful to credit to you as the originator."[55] Langley wrote back to Eddy sometime in October of that year, expressing interest in Eddy's kite experiments.[56] Eddy's experiments prioritized constructing a kite shape that would allow a kite to be raised as high as possible and remain as stable as possible in the air.[57] As Eddy wrote: "I am trying to invent a kite that will remain very long high in the air when its string is released. I have already had hexagon kites at a great height break away and land nearly two miles away. I believe that a kite properly weighted and balanced may go a great many miles before descending at present. I am working at the best shape for this kind of transit. . . . The best results can be attained when the kites are very high, say 1800 feet."[58]

In November of that year, Eddy invited Langley to Seventy-Eighth South Street to "come out and see [my] attempts at kite experiments, or rather the shapes [I have] tried with an explanation of the results." Eddy also wanted to take Langley back to his house in New Jersey, writing to Langley, "It is only a little jaunt of about thirty minutes to Bergen Point, where there is a pleasant little house on a hilltop where I should be much pleased to make everything pleasant for you. If you remain Saturday, we could go out on the 12:45 p.m. (noon) train Saturday."[59] Langley and Eddy met throughout the early years of the 1890s, and Eddy continued his experiments with kite height and kite design. In a November 4, 1893, letter to Langley, Eddy explained a new experiment he had performed in Bayonne "after leaving [you] at Brevoort house." Eddy went home to his house on a hill and at about 3:00 p.m. he tried an "entirely new experiment": he "sent up a kite in the persistence of rain." The height Eddy reached in the rain was 591 feet, and because it was his "first experiment in the rain" he considered it "rather successful."[60] Eddy used this experiment to look forward, writing in the same letter to Langley that "the wet bulb temperatures at a height of 2000 feet in the rain will be very interesting to the meteorologist as I think little is known of the conditions pertaining to the upper part of a cyclone during precipitation."[61]

Eddy cycled through various funding and exchange networks to procure materials for his experiments. He was supported with meteorological instruments specially designed for his kites by Blue Hill Meteorological Observatory and instrument maker S. P. Fergusson. In 1894, he attempted to negotiate silk string prices through the Coombs, Crosby, and Eddy Co.[62] In 1893, Eddy was given a fifty-dollar grant from the Smithsonian through the Hodgkins Fund to finance "the study of electrical and other conditions of the upper atmosphere by means of kites."[63] Although Eddy's exchange with Langley began in 1891, Langley and the Smithsonian Institution as a whole did not begin to fund Eddy's projects until around 1894. Eddy wrote: "At the suggestion of Professor S. P. Langley, Secretary of the Smithsonian Institution, I began in April of 1894, to experiment with silk cord, with a view of reaching an altitude of 10,000 feet."[64] In addition, the head of the Weather Bureau at that time, Cleveland Abbe, was also passionate about "employ[ing] it [the kite] for meteorological work."[65] With the support of Abbe, Langley, and Rotch, Eddy became a leader in the field of aerial experimentation, garnering continued financial support from the Smithsonian and Blue Hill Meteorological Observatory, and employment at the Weather Bureau under Abbe, "reporting atmospheric conditions at different levels."[66] Eddy was canonized as a national hero in local papers for having "devoted his life to the study of kite flying."[67]

Collaborations between Eddy and members of the Weather Bureau caught the attention of the War Department, in no small part because Eddy was often positioning the use of his kites in relation to military reconnaissance. Unlike Nadar, who refused to cooperate with intelligence services when they approached him with regard to aerial surveillance, Eddy was ready and willing to serve the settler state as a meteorological-military kite expert. In July 1897, Eddy and Lieutenant Wise experimented at Governors Island in New York City with "display lights" on elevated kites for the purposes of military and naval signaling.[68] In an 1898 article, "The Naval and Military Use of Kite Photography in Time of War," Eddy discussed how his kites could be used to survey militarized zones: "In the navy the five cameras can take the entire horizon line at sea, giving views of the vessels below the horizon. . . . It is obvious that the location of an enemy's fleet would be revealed, since at a height of 900 feet the horizon receded to a distance of forty miles not taking into account the refraction of the light." The secretary of the Smithsonian, Samuel P. Langley, who collaborated extensively with Eddy, was asked to assist military commissioners in 1899 to develop a "Flying Machine" for the purpose of acting as an "engine of destruction in wartime."[69]

Aerial experimentation and data collection allowed settlers—practitioners like Eddy and reading publics of *Scientific American* and

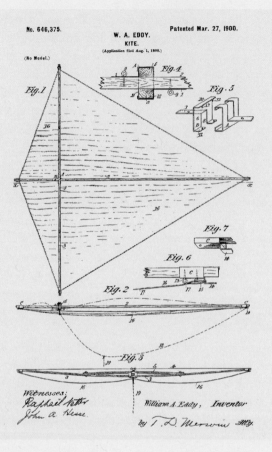

No. 646,375.   Patented Mar. 27, 1900.

W. A. EDDY.

KITE.

(Application filed Aug. 1, 1898.)

(No Model.)

FIGURE 3.1. Patented Eddy Kite. W. A. Eddy. 1900. United States patent no. 646,375, filed August 1, 1898, and issued March 27, 1900.

the *New York Times*—to develop a new "air-mindedness," "a technics of vision."[70] For settlers, developing this new air-mindedness meant doing so through a relationship with the War Department's military technology. Settler experimenters—like Eddy—not only impacted the origins and evolution of popular upper-air data collection but also sought to aid in quantifying the air by sharing technologies with the War Department. Eddy patented the "Eddy Kite" (figure 3.1), and according to *The Photo Miniature*, his kites were sent to Puerto Rico during the Spanish-American War to aid with military photography at "great altitudes."[71]

Scientific discovery, wartime technology, and settler futurity were deeply intertwined, and Eddy's travels between professional communities (like the Smithsonian Institution and Blue Hill Meteorological Observatory) and federal infrastructures (like the War Department and the Weather Bureau) show the ways in which individual experimenters acted as engines of the nation-state. Eddy's popularity in newspapers and scientific journals such as the *New York Times*, the *Sun*, *Scientific American*, and the *Meteorological Review* allowed him and his work to gain popularity among the masses. Such popularity posited data as a public settler good deeply aligned with quantifying space—from land to air—and nation-building. Eddy, daytime accountant and son of an Illinois Baptist clergyman, was a bridge between nondominant communities of settlers and dominant settler organizations like the War Department.[72]

While Eddy was collaborating with the War Department at this time, national populations were looking to the air as a valuable meteorological space, if not a value or commodity itself. On June 28, 1896, the *New York Times* published a three-column article titled "New Instruments in Use at the Weather Bureau," which profiled the use of kites in meteorological experiments and weather data collection efforts. The secondary headline read: "Ingenious devices invented and adapted by the officers of the department—C. F. Marvin's work—kites recently built for the purpose of carrying up instruments for determinations of upper air conditions. . . . Even a person interested in the science of weather observation would find himself somewhat at a loss among the many complicated instruments at the Weather Bureau here."[73] The *Times* depicts a meteorological scene that places the "uninitiated public" against emerging kite practice and knowledge—the "complicated instruments" and "bewildering" vocabulary would fail to make sense to "even a person interested in science."

The *Times* takes on the role of educator, outlining the basic physical properties of two particular kites, the Hargrave cellular kite and the Potter diamond-cell kite. The article included a discussion of the wire attached to the kite, as well as sensing and recording mechanisms, including barometer, tipping bucket attachment, and the nephoscope (a tool for measuring the movement and speed of clouds). The writers take care to describe the technical specifications of each instrument—on the Marvin normal barograph, or registering barometer, "a glass tube hangs by a hook to the short arm of the scale"—and its use in space.[74] The level of descriptive and technical detail in this article signals toward the important role print would play in educating an "uninitiated" public. The *Monthly Weather Review*, established in 1871 and edited by Cleveland Abbe—a former volunteer for the Smithsonian Institution—published 324 articles related to meteorological kites and kite data between 1890 and 1900.[75] The information that the public gained regarding weather kites, and the experiments that could be performed with them, provided access to new atmospheric data sets and also produced a newly quantified way of knowing the sky that positioned its value through settler masculinity and territoriality.

In May 1896 *McClure's Magazine*—which went to press in 1894, selling for fifteen cents, with sixty thousand subscribers—published a series of Eddy's photographs alongside an essay that included instructions on how to build and fly a meteorological kite.[76] *McClure's* publication of Eddy's photos is important because it emphasizes photography's value as a form of evidence for scientists and naturalists. Rather than focus on the artfulness of the photograph, *McClure's* focused on the quantitative possibilities inherent in aerial views. The photographs would be "of the greatest value to geologists, mountain climbers, surveyors, and explorers" because they made visible "wide-stretching landscapes and extensive cities" and "panoramas of every description."[77] Aerial views provided "wide" and "extensive" "descriptions" that could be used by climbers and geologists. In short, the photographs were framed as quantitative records of the earth meant to be studied by a wide public.

Eddy understood his photographs in a similar way, writing in 1900 that with a kite, the camera "can depict the circumjacent earth and its surroundings." And later, echoing his interest in utilizing photographs for military purposes: "For purposes of scouting in military operations the kite has demonstrated absolute and great value. By its means the camera can be sent to a height of 4,000 feet, and made to take pictures of the surrounding country. In this manner, I have taken pictures of New York, Boston, Philadelphia, Washington, and other cities, getting excellent results."[78] Eddy's focus here is on the data gained through such "depict[ions]," even or especially when they aided in military surveillance. Eddy's weather-kite camera invention meant that the settler state—from nongovernment to government populations—could now look down on its weathered terrain from a wholly new perspective. As a form of settler experimentation, weather kites fostered a settler subjectivity rooted in national science as a form of nation-state belonging.

Eddy's experiments gained popularity, and so, too, did the topic of meteorological kites. This topic grew in popularity in scientific journals, national newspapers, and amateur guidebooks across the 1890s. The same year that Eddy began his camera-kite experiments, the *Monthly Weather Review* published an essay on weather kites, titled "A Weather Bureau Kite," under the section "Special Contributions."[79] Professor C. F.

Marvin, author of the essay, begins the work by imagining his reading public who will, by building their own kites, help the "boy's kite" "serve a useful purpose." Marvin continues: "There will probably be many observers who will be delighted and interested to possess and fly one of these seemingly oddshaped cellular kites." [80] In anticipation of a public network of observers and weather enthusiasts who would be "delighted" to build their own kites and run their own cellular kite data collection experiments, the essay is a compilation of written and visual instructions for building and flying a cellular kite. The task, as Marvin outlines it, is for casual observers to turn a boy's toy into a scientific tool. From white elites to white experimenters, and now to settler publics, kites occupied the imaginations of experts and amateurs—it was the tool that tied them together and helped them imagine a vast nation bounded by a sky of data and kites.

The nontechnical language that Marvin uses to describe the kites, in addition to the essay's focus on kite construction, imagines a public of "many observers" who have little technical expertise but a lot of weather enthusiasm (and perhaps many "toys" in their attics). Marvin writes: "The one [kite] of which a detailed description is given in this article is selected for the reason that it is among the best, and at the same time is not very difficult to make." [81] Careful not to overwhelm a public that may have lacked the technical and linguistic expertise that made organizations like the Weather Bureau itself possible, Marvin provides step-by-step instructions for assembly, taking time to outline the most effective materials one can use in construction. Marvin not only assures his readers that this is an easy device to build but also provides instructions for building and use itself. No detail is spared; each description is "given with great minuteness." The instructions read like a recipe. "It is very important that the wood used for the sticks be light and straight grained," Marvin writes. And: "Soft white pine is probably the best. . . . Spruce is stronger but more difficult to procure." [82] As Marvin's essay shows, the passing of information from one experimenter to another (such as from Marvin to his reading masses) meant passing on experiences of success and failure: opt for spruce if you can; always use lighter grades of cotton.

Settler experimenters, professional scientists, and government officials in the United States were quantifying the air, and they relied on print publications to do so. Sketch artists joined these efforts, providing visual representations of meteorological kites. These representations depicted kites floating on the white space of the page as purely technical

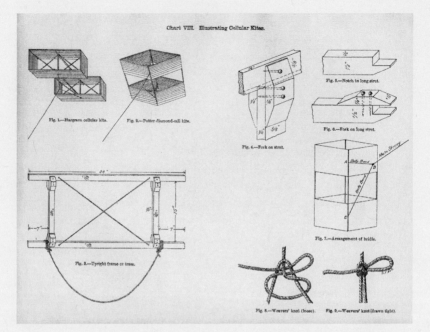

FIGURE 3.2. "Illustrating Cellular Kites." From Charles F. Marvin, "A Weather Bureau Kite," *Monthly Weather Review* 23, no. 11 (1896): 418–20.

instruments. Along with linguistic explanations, drawings and sketches of weather kites proliferated throughout the 1890s, and Marvin's *Monthly Weather Review* essay from 1896 includes a diagrammatic sketch of various parts of the kite. "Illustrating Cellular Kites" appeared as an appended item to Marvin's "A Weather Bureau Kite" (figure 3.2). These visual guides helped spread technical meteorological kite expertise among experimenters, who were imagined to be "delighted and interested to possess and fly" one of them. At the beginning of the 1890s, when settler experimenters had begun tinkering with weather kites, their most urgent questions included how to raise the kite as high as possible. As the century moved on, both elite and amateur experiments were attempting to record temperature and barometer readings while in the air. By the end of the century, meteorological kites had entered the mass print space of popular science journals and magazines as a masculine tool, a device for collecting data and engaging in war. As settler weather data cultural practices shifted to incorporate additional tools and techniques for measuring the air, print culture supported the formation of settler-defined meteorological knowledge. This settler literacy would prove crucial in the century to come.

While Eddy's photographs were not always contextualized as meteorological matter—they did not help to visualize clouds or hurricanes, for example—they were a distinct and important outgrowth of military-meteorological experimentation. This means that while in a practical sense, weather kites (aided by the camera) changed the physical perspective through which everyday people were viewing landscapes, on another level, weather kites also provided an opportunity for experimentation with a variety of technologies not directly related to the study of weather—aerial photography and wireless telegraphy.[83] These technologies found a temporary home on weather kites and flourished there. After weather kites, the US military would begin attaching cameras to rockets, and later airplanes. While technologies like the camera and wireless telegraphy would flourish beyond the kite itself, they would also drastically alter the transmission mechanisms and imaging capabilities of meteorological data, highlighting a need for greater imaging capacities and efficient data transmission systems.

Where in the earlier part of the century institutions and volunteers turned solely to numbers to make sense of the worlds they claimed, late-century experimenters engaged photography and graphical representation through automatic data collection to display quantified environments. Weather kites encouraged settler relationships to data in the late nineteenth century that were optical in nature, rather than being strictly numerical as they had earlier in the century. While numerical weather tables would persist throughout internal Weather Bureau publications, new roles had been defined for data collectors and new visual standards had been codified for consuming publics.

TOWARD AUTOMATED DATA

Throughout the nineteenth century there was a dynamic exchange between lexical, numerical, and pictorial weather data displays that allowed for settlers to develop an environmental consciousness bound to science as nationalism. In the early to the middle part of the century, lexical displays dominated print meteorology. Slowly, numerical and specifically tabular displays gained popularity, and by 1855 they had begun to dominate the weather data display scene. Grounded in an interplay between numerically based instruments, human observation, and recording, tabular displays became interlinked with the very notion of weather as data for settler populations at the time. But by the end of the nineteenth century, the nature of the dynamic between lexical, numerical, and newly pictorial display underwent a significant shift. In the early to middle part of the century, science was a form of nation-building, where settler colonial expansion across the eastern settler states necessitated meteorological science. The data tables produced in this period became part of settlers' environmental legacy, or how they narrated their relation to land. With the rise of aerial weather data collection through kites and remote-sensing instruments, weather served as a cultural formation imbued with these histories of settlement and violence. Narrating the past, and imaging a future, of the US settler states relied on older tabular and newer forms of optical meteorological data.

Throughout the late nineteenth century, no single party was in control of upper-air imaging, data collection, and weather tool construction. Official efforts at the Weather Bureau existed alongside and were often influenced by experimenters. Observatories such as Blue Hill Meteorological Observatory and other government sectors such as the War Department would join this effort, utilizing knowledge from experimenters to fuel their own meteorological kite endeavors. Print publications helped foster technical expertise among settler experimenters because such publications often included extensive construction and use guidelines (while also articulating a growing distinction), as the 1896 *Times* article makes clear, between professional meteorologists and backyard experimenters.

Several different parties were experimenting with automated data collection. At the War Department—which headed the National Weather Service from 1870 to 1890—automatic charts and automated weather data collection experiments produced new ways of recording wind, atmospheric pressure, and sunshine through a variety of self-registering instruments, from self-registering anemometers meant to record wind readings over a twenty-four-hour period to automatic sunshine recorders. Daily records were bound into yearly books that formed a graphical catalog of wind lines and sunshine imprints.[84] This way of rendering wind and sun was notably different from the ways that settler observers and army personnel had learned to render wind through numerical standards from the early nineteenth century up until 1870.[85] Automatic instrument recordings produced distinct data records that communicated in the language of graphics rather than the language of numbers. The War Department would continue with these experimental data-gathering and data-recording practices throughout the 1870s, including continued experiments with photographic sunshine and with self-registering barographs.

Between 1894 and 1895, Eddy—along with instrument maker S. P. Fergusson—raised a self-recording thermograph and meteorograph on a kite in Milton, Massachusetts.[86] On September 15, 1895, the *New York Times* published "William A. Eddy's Kite Experiments," which showcased Eddy having lifted a thermograph to the height of 1,916 feet, "probably the highest altitude in the world."[87] By the end of the century, the federal government had developed seventeen kite stations across the Mississippi Valley equipped with kites that could produce automatic charts at the height of a mile.[88]

Self-registering instruments produced graphical products similar to, but substantially different from, the numerical tables that grounded previous generations of settler observers. These graphics encouraged a way of conceiving of the weather that relied on imaging and optical translation. At the Smithsonian and elsewhere (including popular and scientific publications such as the *New York Times* and *Scientific American*), weather took on a lithographic, photographic, and aesthetic quality at the very moment that upper-air exploration through kites and balloons took off. While weather table publication would persist well beyond the transfer of the Smithsonian meteorological project to the War Department in 1870, the Smithsonian and other weather publication venues would begin to publish weather materials that favored a pictorial weather sensibility rather than an exclusively numerical one. This enabled a unique environmental consciousness—one grounded in graphics—to emerge.[89] Throughout its ownership of the weather project, the War Department continued to employ station agents to record instrument-based recordings in tabular form at newly established weather stations across the country. But the War Department and Weather Bureau did not utilize volunteers in the same way. While they maintained a network of volunteers retitled as "cooperative observers" throughout the twentieth century, they hired employees with the sole purpose of managing the meteorological data collected and registered by automatic recording instruments.

Weather maps were directly dependent on settler data, and they were used to characterize "American" weather by visualizing the territory of the nation-state. By 1891, twice-daily weather maps gathered data from 160 places across the United States and covered an area of more than 3.5 million square miles, "about one-fortieth part of the surface of the whole globe."[90] Daily data was communicated to the central Weather Bureau office in Washington, DC, through telegraph "with the aid of a cipher code."[91] A "skilled translator" in the Weather Bureau office translated the cipher codes back to data, and an "assistant enter[ed] the data upon a map of the country." The data was then "generalized by lines representing graphically the various conditions of the air and the change going on."[92] The map was issued two times a day in 1891 in post offices and customhouses, boards of trade, and maritime exchanges across "a number of large cities throughout the country." The bureau recorded that over 5,000 copies of the map were issued daily in 63 cities at a size of 22 by 16 inches.[93] Mapping not only turned numerical data pictorial but also unified the various settler territories into the "United States," a land mass bound by weather data and the histories of settlement and violence it contained. What was countable could be quantified; what was quantifiable could be known. Espy's "wide extended country" had been achieved.

From the 1890s on and into the following century, aerial meteorological data collection remained a cornerstone practice in American meteorology, from kites to ballooning and rocket photography to commercial jet weather data collection and satellites. All the while, optical imaging and remote meteorological sensing also remained linked. This linkage shaped settler relationships to data, the environments they inhabited, and the lands they claimed as their own. While the popularity of the meteorological kites as a dominant aerial technology would fade throughout the early part of the twentieth century—replaced first by balloons and later by rocketry—the notion of the self-registering instrument would not. In fact, self-registering instruments would become a mainstay of meteorological data collection across US Weather Bureau stations in the early to mid-twentieth century. This is where the next chapter takes off: an inquiry into the life of self-registering instruments and the ways in which early twentieth-century environmental data culture was marked by automated technologies, which aided in developing an environmental consciousness grounded in expansion, domination, and environmental control.

4

DATA'S EDGE: CLEANING DATA AND DUST BOWL CRISES

In rooms 304, 305, and 322 at 102½ Spring Street in Los Angeles, California, local forecaster George Franklin consulted his compilation of weather measurement instruments.[1] Spread across three observing rooms, Franklin cared for and transcribed instrument-measured data from a barograph, thermograph, sunshine recorder, and self-recording rain gauge.[2] He noted the maximum and minimum thermometer readings from the self-registering thermometer and transcribed each reading in black ink onto form 1001. On June 6, 1900, Franklin noted a high of 88.9 degrees and a low of 53.6 degrees. Beside each daily entry, Franklin signed the column titled "Initial of Observer" with his initials, GEF. For in acting as an observer, Franklin was obligated to make sure that every meteorological observation was "verified each day." By signing his name, he acknowledged he was responsible for the "accuracy of the form" and, more obviously, the accuracy of the data that form displayed.[3]

Twice daily, at 8:00 a.m. and 8:00 p.m., Franklin would walk to the Western Union telegraph office—past Hotel Nadeau and the county courthouse—in downtown Los Angeles to file his daily meteorological form.[4] Form no. 1001-Met'l was an eight-page form—used by Franklin and others—on which monthly weather data from automatic instruments was recorded and held. General instructions on the form itself read, "This form will be kept clean, and all entries therein made in a neat, legible hand." Several additional items follow under these general instructions, including that the leaves of the form should be separated for "convenience in entering the data" and that "all entries in this form must be made in black record ink," as "red ink will not be used except in underscoring." The instructions note that if an error is made, the erroneous data should be stricken and the correct data should appear above the strike in "black ink."[5]

Form 1001 and the three weather station rooms at 102½ South Spring Street were material manifestations of the settler science that had been imagined and enacted throughout the nineteenth century: Franklin's office was a node in James Espy's "wide extended country." This science and the worldviews it coded within itself were also transforming at the turn of the century—a transformation that is the subject of this chapter. In the Weather Bureau's inaugural 1890 report, its chief wrote that "the great tendency of the genius of modern times has been to devise and invent instruments and mechanisms that shall not only do all manner of work, but also very often do it much better than could be done by hand."[6] The desire for instrument-based labor was tied to desires for continuous streams of data. The bureau also wrote that "to ascertain this climate we must have continuous records hour after hour and day after day of all the meteorological elements." Franklin had a recording station that could keep "continuous records hour after hour." By 1900, the bureau had successfully incorporated "ingenious instruments" that captured and visualized data in historically new ways. Stations across the nation were equipped with more automated recording instruments as well as storm-warning signaling devices, including oil-burning and electric storm-warning lanterns.[7]

The nineteenth and early twentieth centuries brought about a kind of statistical thinking, as Theodore Porter has shown, that encouraged the evolution of standardization.[8] This was especially true at the Weather Bureau, as standardization intensified over the course of the late nineteenth and early twentieth centuries. Scholars Andrea Rusnock, Matthias Heymann, and Lorraine Daston have signaled toward the dynamic history of standardization in the visual history of data tables themselves, a deep concern of this book's first chapter.[9] Almost a century after the first US meteorological forms circulated in upstate New York, general instructions and manuals had come to articulate a discourse for eliminating and correcting data errors. In the early twentieth century, this discourse grew in meteorological forms and Weather Bureau instruction books, pressing the distinction between "dirty" and "clean," "erroneous" and "error-free" data. This discourse was the life force of data standardization itself in the post-1890, pre–World War I period. Bureau officials did not have patience for what could not be quantified, or what was quantified in error: the edges of data signaled toward an unknowable world that, from the perspective of the Weather Bureau, would not be tolerated.[10]

Thus far, this book has studied the ways in which settlers historically utilized weather data to achieve certain US settler state ends. As chapter 1 explored, in the early part of the nineteenth century, settler claims to land were closely aligned with, and necessitated, weather data collection. Chapter 2 investigated how, in the mid-nineteenth century, weather data culture helped to build a national meteorological practice that bolstered an environmental consciousness founded on heteropatriarchal science. Chapter 3 addressed the ways in which this environmental consciousness was furthered in the late nineteenth century through male-dominated upper-air experimentation and data collection. Across all these chapters, I have shown how meteorological data helped to enact settler presents and develop settler environmental domination of biospheric life. Settler presents were defined by a colonialism that, as Max Liboiron has argued, "means ongoing settler access to Land for settler goals," which "includes access to futures."[11] Across the nineteenth century, weather data culture assumed access to land for collection and processing. The early twentieth century continued and intensified this trend, as continuous collection and collation of data was established as a norm. This norm also suggested a future where data could be easily stored and processed—this, too, assumed land. In support of such a present and future, weather data discourse shifted toward "messy" and "clean" data.[12] This shift was important because it allowed for greater efficiency: when marked as clean or dirty, data could be easily controlled—divided into categories, discarded, or heralded as essential.

Although the rise of professional meteorology shaped the history of meteorological science in valuable ways, this chapter does not focus on professional meteorology. Instead, I explore how Weather Bureau settler data culture—narratives about good and bad, clean and dirty data—

informed the environmental consciousnesses of almost six thousand Weather Bureau cooperative observers and employees, as well as the environmental consciousnesses of settler publics receiving data reports in the form of storm warnings and forecasts across this period. While the rise of professional meteorology shaped what it meant to be a meteorologist (allowing meteorology to become a theoretically driven atmospheric science) in the post–World War I period, both before and after the war, the Weather Bureau remained an important site of data production that continued to shape US settler state data science and sentiment.[13]

In the early twentieth century, the Weather Bureau maintained an extensive network of data collection points, many inherited from the Smithsonian Meteorological Project.[14] In 1909, the Bureau had 1,915 employees outside of Washington, DC, who worked in roles such as river, rainfall, and special meteorological observers; it also had 4,695 cooperative observers.[15] Despite its extensive data collection network, the Weather Bureau was not able to develop active research agendas. It was mired, instead, in a need to "produce forecasts and warnings." Collected data was aimed toward practical, day-to-day forecasting, and so applying "mathematical and physical principles to the data" and experimenting with theoretical research on the atmosphere and long-range forecasting was not possible.[16] Clean data fostered efficiency as the Weather Bureau worked toward its end goal of day-to-day forecasting for the nation.

At the Weather Bureau, instrument-oriented and automated collection helped to construct a relation to clean and dirty data, while also encouraging relationships with weather itself, rooted in continuous data access and data continuity. At the same time, data became foundational for observational forecasting. Franklin and his contemporaries were tasked with integrating technologies such as automatic and self-registering instruments into daily weather data collection and, at times, utilizing local data for weather forecasting as "local forecast officials," a title introduced in 1894. While modern-day forecasting emerged in the US interwar period with the proliferation of calculating aids, punch card machines, and mapmaking, the role of forecasting in the prewar period was dependent on observational data rather than theoretical or equation-based models.[17] Data needed to be clean and error-free for the historical record, but also to ensure accurate forecasting.

The first section of this chapter focuses on the Weather Bureau and contends that data evolved there through discourses rooted in fears of mismanagement—errors in recording or transmitting data—in addition to fantasies about optimally transferred and clean data. I explore how land-based and wireless telegraphy formed foundational transit pathways for the circulation of meteorological data. I then examine how instruction manuals and Weather Bureau reports framed data care through a rhetoric of management that positioned cleanliness as an ideal and dirtiness as a fear. Bureau officials were responsible for ensuring that data was clean and good, rather than dirty and bad, and the emergence of a qualitative discourse to assess quantitative data marked an important shift in data practice. It marked a disciplining of environmental data capture and culture, an "enumerative strategy" that buttressed and created a coherent colonial space through good and clean data.[18] With self-registering instruments and automation, Franklin and his contemporaries were forging distinct relationships with meteorological data that enacted fantasies about the weather—namely, that is could be endlessly quantified through categorical separation—and national science.

As data landscapes grew more comprehensive through self-registering instruments and twenty-four-hour data streams between 1900 and 1930, bureau employees and volunteers grew further estranged from the environmental worlds that they had been measuring for over a century. This estrangement led to a data crisis during the Dust Bowl of the 1930s—the subject of the chapter's second section—a crisis in which measurement and management regimes failed to register the escalating catastrophe of airborne dust and soil erosion. The ultimate "dirty" data, dust fell out of data-capturing systems and could not be easily incorporated into existing measurement regimes. Existing systems of automation and self-recording—and quantification more generally—failed during the Dust Bowl, while hand-based data collection practices prevailed, challenging the accepted modes of data collection, collation, and visualization that had developed since the early nineteenth century.[19] Observers failed to make sense of the dust crisis on two distinct levels—they not only were unable to measure the crisis but also were unable to make sense of the reciprocal ties that bound them to the very crisis they were measuring.

The dream of a continuous, automatic, standardized, and autonomous weather data was a vital part of weather data culture in the early to mid-twentieth century. Facilitated through the advancements of land-based and wireless telegraphy, such dreams aligned settler desires for environmental control with a future of endless data, or, in the words of Elena Aronova, the "data regime" of the 1950s and 1960s, where planetary geophysical data became an accumulated and stockpiled "a form of currency."[20] The crisis of the Dust Bowl spurred this data regime as well, as *more* data was poised as a solution to the unfolding crisis in the West. But at the root of the Dust Bowl crisis was a way of relating, and that way of relating would only be sharpened by access to more data.

TRANSMITTING DATA

Before data emerged as dirty or clean, Weather Bureau officials had to figure out how increasingly large streams of data would move across the nation. In 1900, the Weather Bureau hired wireless experimenter Reginald Aubrey Fessenden to run wireless telegraphy experiments at Cobb Island outside of Washington, DC.[21] These experiments were so successful that the Weather Bureau established additional wireless stations in North Carolina and Virginia. The bureau's 1900 report read that the secretary of agriculture "has authorized the Weather Bureau to systematically investigate the various methods of electrical communication without wires" and that "the progress already made in this investigation is eminently satisfactory." And, in 1901, the bureau remarked that "the line of research had been divided into three classes": first, perfection of a "more powerful transmitter"; second, "the devising of a more delicate receiver"; and, third, finalizing a system of "selective telegraphy" where only the "receiver desired" receives the telegraph message.[22]

Wireless telegraphy introduced the possibility for the Weather Bureau to receive "weather information from places where there were no telegraph wires, telephones or cables, and from foreign countries and ships at sea."[23] While the implementation of a wireless telegraphy system would have allowed for greater data transfer, it would be quite some time before wireless telegraphy could be integrated into the observing system. As the Weather Bureau continued its experiments, other sectors of the federal government began to lay claim to its experiments and inventions. In 1904, the federal government convened an interdepartmental board, appointed by the president, to "consider the entire question of wireless telegraphy in the service of the national government." What the board found was that wireless telegraphy would "serve" the navy more efficiently than the Weather Bureau, which was housed within the Department of Agriculture. The board required that the Weather Bureau turn over all coastwise wireless telegraph stations and meteorological data to the Navy Department "at no cost to them."[24] While the Weather Bureau needed to turn over its coastwise stations to the navy, they would be granted the right to receive meteorological observations once daily through naval wireless stations and were able to communicate storm warnings through these stations as well.[25] One thing was clear—advances in wireless telegraphy belonged first to the navy and only secondarily to the Weather Bureau.

Because the navy now owned and operated wireless telegraph stations, this bound data transmittal to daily mail, railway telegraph daily transactions, and telephone forecasts and warnings in the opening years of the century. These would remain the dominant modes for weather data transmittal. In 1904, New York alone distributed 305 daily railway telegraphs, 961 telephone special warnings, and 6,985 pieces of mail. In 1907, the bureau wireless telegraphy division was entirely composed of the collection of meteorological observations "from vessels at sea, and the dispatch by the same means to vessels at sea of weather forecasts and storm warnings based upon the observations thus collected."[26] Seven different lines, all equipped with a Marconi apparatus, had been authorized to transmit data to the bureau that year. The bureau continued to develop wireless capacity in the only way the federal government would allow—though its partnership with navy vessels. By 1912, the bureau had equipped more than thirty vessels with wireless telegraphy, with all of this equipment owned and operated by the navy. Vessel weather service helped to expand data collection to ocean- and coast-based stations, and ultimately expanded the data that could be used for regional and national forecasting. At the same time, it created distinctly new data streams that needed to be managed and "cleaned."

By 1915, an emerging wireless system—one that allowed for the broadcast of weather forecasts and storm warnings, and for radio users to tune in to vital weather data—overlaid land-based telegraph systems and postal systems. Fifteen years after their first experiments outside of Washington DC, amateur wireless operators employed by the Weather Bureau implemented a land-based wireless weather forecast distribution that did not need to pass through navy control. In June 1915, a Weather Bureau station in Illinois distributed wireless weather forecasts to wireless operators in the region. In University, North Dakota, a similar practice was underway as early as 1914. The Weather Bureau had wanted to initiate a wireless data transfer system since 1900, but it was increasingly difficult to do with limited funds. Land-based telegraph lines were often affected by extreme weather—the subject of the communication threatened to annihilate the very system itself. "Almost the first effect of great floods and destructive storms," the bureau wrote in 1915, "is to cut off communication by the customary wire service." But "wireless methods of communication are subject to but little, if any, interruption by destructive weather conditions. . . . A powerful argument is found in these considerations for the establishment of wireless stations in many regions of the country, especially those that have repeatedly suffered from disastrous floods and storms and the serious loss of communication with the outside world."[27] By 1916, wireless service had been "somewhat extended," with forecasts for nine states "distributed from four points." These forecasts were received at 270 amateur radio stations.[28]

Land-based telegraphy, the postal service, and wireless technology were intimately tied, as they formed foundational transit pathways for the raw observational data around which forecasts were built. The transmission of timely and readable observational data was necessary for station weather maps, as well as a variety of forecast products, some of which were broadcast via wireless telegraphy. In 1916 the bureau printed over 7 million station base maps, over 400,000 daily forecast cards, and over 7,000 weekly forecast cards. At this time, 199 principal stations and 4,500 supplementary stations or substations were moving weather data between land-line telegraph stations, wireless telegraph stations, and telephone lines.[29] As data moved between land-based and newly emerging wireless systems, a new discourse of care emerged for the weather observer and the station agent. These discourses moved observers from the role of data collector to the role of data caretaker, which also required passing "competitive examinations."[30] Once a practice centered around the sensing body—as Espy encouraged his observers to "keep the head steady in one place"—gathering observational data was now centered around the life force of instrumentation.[31] Station agents became caretakers of weather data and weather instruments, rather than primary producers of it.

The underside of care was error. But what did erroneous data look like, and were there trends in who produced this bad data? Across the late nineteenth and early twentieth century, the Weather Bureau articulated, in a variety of ways, that error would not be tolerated, emphasizing clear data values such as verifiability and error-free recording and reduction. Meteorological form 1001 is one important example of how the Weather Bureau communicated its values around "clean" and "error-free" data. But, in addition to forms, the bureau utilized instruction manuals and employee exams to articulate and enforce clean data collection, calculation, and transcription. Overtly valuing clean data allowed the bureau to redefine its requirements for employment. In the Department of Agriculture's annual report for 1900, the Weather Bureau explained its guidelines for "a system of merit and discipline."[32] In articulating this system, the bureau developed strict standards for admittance and continuation as an employee. Both were uniquely tied to the production of clean meteorological data.

Systems of merit—instituted in 1899 through "Instructions No. 79" and communicated through instruction manuals and employee regulations—were used to develop and enforce standards for clean data. An observer had to "prove himself" in a variety of subjects, including arithmetic, elementary physics, algebra, and advanced meteorology, and "possess certain scholastic requirements, including a knowledge of the science of meteorology," in order to act as an observer.[33] Admittance to the Weather Bureau was no longer based simply on a desire to practice meteorological observation but rather was dependent on passing a series of entrance exams and engaging education offered by Weather Bureau associates.[34] While meteorology did not become a professional science until the interwar period, admittance standards had evolved to include examination.[35] Once admittance into the ranks of observer was achieved, observers needed to adhere to particular station regulations.

In 1905, Willis L. Moore articulated the "Regulations for Conduct of Stations," a set of "regulations, instructions, and order" that were adopted by the Senate in 1907.[36] Much like the general instructions, these regulations emphasized accuracy and verification: "Each observation should be verified when recorded, and all entries, compilations, and computations of data must be checked each day by the employee responsible for their accuracy."[37] Efficiency records were kept for each employee. These records reported the "accuracy of each employee" and were maintained and held at the Weather Bureau office. There were rewards for accuracy and punishments for errors: employees who made no more than nine errors in six months had "that fact noted on their efficiency records"; employees who made more than nine errors were subject to "investigation" into their "fitness for continuance in the Bureau," "admonishment," and "disciplinary action."[38] Implementing a merit system not only reduced the general number of Weather Bureau employees acting as observers and local forecasters but also reduced the particular number of self-identified women observers who were part of the bureau's employee force.

In the early part of the twentieth century, calls for data accuracy and clean data evolved in tandem with a remasculinizing of the Weather

Bureau, eliminating women observers from the ranks through "competitive examinations." In 1903, over 350 people were acting in the roles of observer, assistant observer, or local forecaster at the Weather Bureau. But only one self-identified as a woman—Mrs. Margaret E. Conway.[39] Student assistants were all male, and one additional self-identified woman—Mrs. Elizabeth Renoe—was noted as a station agent.[40] Self-identified women are indeed found under the "Cotton Region Observer," "Corn and wheat region observer," "Special river," "Fruit and wheat region observer," and the "Storm warning display man" roles. However, these roles did not require continuous observation, as, for example, corn and wheat region observers were asked for only one observation per day, from April 1 to September 30. Ensuring the accuracy of data through a merit-based system produced, by default, a workforce of more than 350 forecasters, observers, assistant observers, and station agents that included only 2 women but more than 30 individuals named "William" in upper-level bureau positions.

While women were omitted from station worker employee counts in the early part of the century, their service was often defined through organizing categories that signaled male work and advancement.[41] "Weather Bureau Men" were "educators"; "Weather Bureau Men" were "University Students."[42] If women were interested in climbing the ranks at the Weather Bureau, they would have to contend with stated preferences for male work. A 1920 ad for an assistant observer read: "All citizens of the United States who meet the requirements, both men and women, may enter this examination," but "for this position in the Weather Bureau male eligibles are desired."[43] Data practice continued to evolve as a masculine practice in the early-century period, and opportunities for data caretaking privileged the white male body as an ideal. At the same time, approaches to data care—from collection and reduction to instrument care—were built and shaped by male hands. Women did maintain a presence as cooperative observers, though they served in this force without compensation.[44] In 1912, "a few women" were reported among the cooperative force in Utah, including "[a] lady on a far-away ranch in a southeastern county some 50 miles from a railroad." The *Monthly Weather Review* reported that "their records have made every appearance of being made with all the care and precision possible."[45] While women maintained a presence in this force, they did not rise to the ranks of local forecast official, observer, or station agent at the same rates as men did throughout the early-century period. Examinations for promotion were built around testing in arithmetic, English, grammar, and elementary meteorology for promotion to the $1,000 grade; algebra, elementary physics, and plane trigonometry to the $1,200 grade; and astronomy, plant physiology, and advanced meteorology for the $1,400 grade. The education that young women were receiving and pursuing from elementary to high school did not set them up for success in these exam areas.[46] While the nature-study movement had opened scientific study to generations of women, shifts in education in the early twentieth century interrupted this developing path.[47] A national weather system grew in tandem with a national data politics that prioritized heteropatriarchal data care.

A 1915 publication, *Weather Bureau Topics and Personnel*, a monthly pamphlet of news and helpful advice meant for the mostly male station agents scattered across the country, documents the daily life of station agents and especially their interactions with instruments and data. In the monthly repeating section "Care of Instruments," we see that station agents were not just asked to compile daily weather maps by stencil, chalk, and milliograph; they were also directly responsible for the care and effective production of self-registering instruments and the graphics these instruments would produce. Station agents were instructed in various topics, from how to control a wind velocity pen when it was abnormally dotting, to securing thermograph and barograph sheets with rubber bands, to replacing the muslin cover on a wet bulb "in large cities, particularly where numerous factories are in operation" and the cloth becomes "soiled and dirty."[48] Station officials like George Franklin, for example, had the daily responsibility of mapmaking by stencil. On stenciling in warm weather, the bureau recommended that station employees "rub a large piece of ice over the ink pad" to "cool both the ink and the pad": "100 to 200 good maps can be run off without reinking."[49] In winter weather, the bureau recommended the use of "hot water" on the ink and pad to loosen up materials that are "often too cold for work." The mapmaking process was also complicated by the length of time required for the ink to dry. While the ink was expected to "dry quickly on paper," the bureau recommended an aid in this process through the use of "Ivory soap, shaved thin and worked into a jelly."[50] The jelly was worked into the ink pad itself and aided in quickly drying the ink on the paper. Station agents were also engaged in chalk plate map work and milliograph work, for which the bureau also provided detailed suggestions for use.[51] From caring for the wet bulb thermometer to rubber-banding recording paper and self-registering instruments to stenciling isobars and isotherms onto national maps, one can see a reconfiguration of the observer with respect to data labor.

Instruction manuals and bureau reports articulated a data code of conduct to largely male workforces in the early twentieth century: an agent's responsibilities could include caring for instruments, learning the Weather Code Book, preparing bulletins and forecast charts, answering phones, and preparing telegraph abstracts. Station agents were sometimes given assistant observers or apprentices who, after advancing within the bureau from "the grade of messenger boy," would learn these tasks through the station agent.[52] Assistant observers were presumed by the bureau to have more technical skill than apprentices and were not allowed to perform observations unless under "especially authorized instances."[53] Not only had the tools for measuring weather grown more dynamic since the turn of the century, but the human networks for recording, transmitting, and caring for weather data had become more dynamic as well. Male observers were no longer the sole sensors of data collection but were caretakers of automated instruments. Station agents, assistant observers, apprentices, and messenger boys were all part of a national network of US Weather Bureau stations, and this allowed people across the nation (from the east to the west and back again) to conceive of weather from coast to coast.

As landscapes for collection, transmission, and forecasting grew complex, the Weather Bureau articulated codes of conduct for station agents that accounted for data loss and incompleteness. In a data landscape where every moment was recordable, there was a new significance given to incomplete records and the "lost" motion of instruments.[54] From the 1890s up until the 1930s, the Weather Bureau reconfigured the very notion of weather data through the distribution and enforcement of instructional manuals and guides that emphasized complete and error-free data, cleanliness, the plainness of data, and "increased vigilance" in making sure that "loss of records or other defects do not occur."[55] The bureau's "Special Instructions" for 1894 emphasized the "great importance of presenting only plain and unaltered facts" in regular observational data as well as obtaining data from automatic instruments and transcribing it into tabular form.[56] Yet, if an automatic instrument failed, station agents were responsible for special observation that would "aid in supplying missing records and help to bridge over interruptions."[57] In "all cases of the failure of the instruments," agents were expected to stand in for what the instrument had failed to do—to supply the missing data, noting "by the use of ink of different color," that this was a backup observation, one conducted by a person rather than by an instrument.[58] Additionally, if "the instrument makes an abnormal or improperly placed record, such portions should not, under any circumstances, be erased from the sheet."[59] The agents' bind was triple: to care for instruments, to transcribe data without error, to note all errors, and to stand in for automatic instruments if they failed to do the work of data collection.

The Weather Bureau emphasized two new values of data practice, which became crucial to the environmental consciousness of the period—"completeness of records" alongside "plain facts." Complete records meant that "errors" or misreadings were to be included in the data transcript. Error and misreading were essential ways in which station agents and bureau personnel configured clean data, or understood landscapes of "plain facts." Completeness meant that "sheets containing imperfect records must never be destroyed," and that if "the instrument makes an abnormal or improperly placed record, such portions should not, under any circumstances, be erased from the sheet."[60] The point was to include all human and instrument recordings, from errors and misreadings to "plain facts." Attention was also paid to the practice of "correction," which encompassed comparisons between standard and automatic instruments. Corrections constituted any difference in measurement between instruments, and these corrections were recorded on correction cards.[61] Corrections cards were used for thermographs and barographs and were registered on graph yards via a rubber band and pin-insert system. Corrections, for example, to the electrical sunshine recorder were determined by "eye observations" registered as an "explanatory note made on the margin of the form and on the original record sheets."[62] Corrections were a shaping force within the data landscape as they constituted the practice of instrument calibration.

With the onset of self-recording instruments, notions of "plain" data, completeness of records, and the practice of correction helped to link station agent data to factuality. The rhetoric of instruction manuals and personnel memos—documents circulating between station agent populations and the bureau itself—show how plain and clean data become synonymous with fact itself.[63] While plainness was linked with fact, neatness allowed for "legible" records. Field station agents were expected to value "neatness" above all in their operations as Weather Bureau employees precisely because neatness was connected to the legibility of data. Neatness was opposed to what the instruction manuals termed "blotted, imperfect, and untidy" records. The value of neatness was located in the care of automatic instruments and the keeping of records themselves. This remained true throughout the life of the instruction book at the Weather Bureau, which was printed by the Government Printing Office in three different editions from 1892 to 1922. The 1913 instructions echoed those from 1894: "When the triple, double, or single registers are in proper condition and adjustment, wind velocities of 90 miles per hour can be recorded without difficulty; but if the pens are neglected and smeared on the outside with ink, dust, and fibers of paper so that lines become heavy and the mile marks merge together at velocities of even 40 miles per hour." The instructions go on to say that this situation can be avoided by "a little attention to neatness," adjusting and cleaning the pen, and caring for the "special" ink and ink pads. Attention is also paid to loading graph sheets, or "forms," onto automatic instruments: "Sheets should be wrapped closely around the cylinder and held by rubber bands." The recorder must be "carefully centered in the appropriate space on the form, so that the respective arms strike over the corresponding letters on the sheet."[64] At stake here is not only preparing the instrument for proper recording (from centering sheets to cleaning ink pens) but also preparing the recordings made by self-registering instruments for shipment to the Weather Bureau office.

Practices of data care emerged from weather instruction manuals, guides, and personnel reports, and these practices positioned the male, observing body at the center of sustaining weather data in the United States. The relationship between automatic instruments and weather data was articulated through the discourses of accuracy and error, plainness of "fact," and the legibility and completeness of the record. Station agents, as caretakers of weather, had the capacity both to corrupt data through faulty processing and to certify data as accurate. While data care had been part of weather data collection from the mid-nineteenth century on, it had never been so clearly aligned with accuracy and articulated as such across as wide a network. By 1911, the bureau employed agents at over seven hundred stations across the United States, and much was at stake for utilization of weather data.

In the meantime, the emergence of the American Meteorological Society (AMS) in 1919 and the onset of World War I deeply impacted the development of weather data production and professional meteorology in the United States. Both of these developments left their marks on the Weather Bureau itself: "Operational needs of the military service during

World War I were responsible for a surge of interest in meteorological support," and such interest "did not subside" after the war was over.[65] Although the Weather Bureau received a $100,000 appropriation to work with the War Department to establish aerological stations utilized during World War I, the bureau remained "underfunded, undermanned, undertrained": "Progress was limited—academically, theoretically, and within the applied sector."[66] Outside of the Weather Bureau, educational opportunities for degrees in professional meteorology grew. Theoretically driven meteorologists began organizing through the AMS and began teaching meteorology at degree-granting institutions. As dust blocked horizons, the Weather Bureau remained in control of national data collection and forecasting for the populations of the US settler states. The bureau and allied government sectors (such as the Department of Agriculture and the Department of Commerce) would face many situations in the years to come—from the Great Depression and the Dust Bowl to World War II and documentation that earth's climate was undergoing historic, anthropocentric shifts—where the veracity of data mattered more than ever.

BAD DATA

On the evening of February 24, 1936, R. P. Haywood of Keene, New Hampshire, noticed that it was snowing outside. The snow transitioned between sleet and rain, and back to snow again, until finally Haywood saw that instead of white flakes, the snow was brown. Haywood collected a sample of the brown snow and sent it to the Bureau of Chemistry and Soils in Washington, DC. The bureau performed a "mechanical analysis" of the snow that resulted in the following facts: Haywood's brown snow was composed of "quartz and feldspar," "opaline and silicified organic remains characteristic of midwestern soils."[67] In 1936, dust traveled from Colorado, New Mexico, Texas, Oklahoma, and Kansas through precipitation and air systems, landing in New Hampshire, Vermont, and northern New York. From Nebraska to New York, from Oklahoma to the Golden Gate Bridge, dust clouds not only crowded fields of vision but also occluded traditional land-based weather measurement regimes, which failed to capture the numerical meanings of dust.

The Dust Bowl of the 1930s was a land-use crisis born out of capitalist approaches to agricultural production, specifically the settler ethos of sodbusting. As dust crowded fields of vision, land-based observational data was not able to offer numerical clarity on the clouded situation. The ways that station agents, official observers, and cooperative observers made sense of hazy horizons reveal that the automatic instrument-based approaches developed in the earlier part of the century were not enough to understand the environments they inhabited and the crisis they had created. Dominant meteorological data capture regimes failed because they were founded on atomistic and heteropatriarchal approaches to understanding the biosphere. These approaches could not account for interlaced earth systems and the human effects on these systems.

Frontier settlement shaped environmental knowing in the region, as shown by the strategic placement of US Army forts—fifteen in Kansas from 1827 to 1885, and nineteen in Texas from 1849 to 1885—that acted simultaneously as meteorological outposts across the region.[68] Kansas and Texas fort-outposts were built on the homelands of the Osage, Kaw, and Comanche Nations, and these fort-outposts facilitated settler expansion by allowing settlers to gain tactical meteorological knowledge that could be used for war and settlement.[69] While the US Army abandoned Fort Gibson in 1890—a fort originally intended to provide aid and defense for settlers as they occupied lands in the nineteenth century and on—frontier settlement and the development of strategic meteorological outposts had indeed been achieved. By 1930, Oklahoma was home to eighty-four stations reporting climatological data for the year.[70] At least two former army forts were still in operation as meteorological posts. Fort Gibson and Fort Reno reported climatological data to the state of Oklahoma throughout the early part of the twentieth century and throughout the Dust Bowl crisis of the 1930s.[71]

As the Weather Bureau continued to nationalize weather data collection through merit and examination—an effort that resulted in a reduced, though "professionalized," workforce that was almost entirely male—it also faced uneven development of official Weather Bureau stations. Official stations tended to mimic the geography of frontier fort settlement, whereas cooperative stations tended to mimic the geography of settler homesteaders. This produced a data collection landscape in which official weather stations were in less prominent dust storm regions, and more cooperative stations were located in the heavily impacted dust storm regions where homesteaders were sodbusting. The western region of Kansas, for example, experienced the greatest impacts of the dust storms, yet there were no official stations in this region. Instead, the region supported a robust network of cooperative weather observers.

Unlike official stations, where data collectors were employees who were paid to take measurements twice daily, or hourly, through a set of self-registering or automatic instruments, cooperative observers were part of the unpaid, volunteer climatological service of the Weather Bureau. In April 1936, there were seven official stations in Kansas located in Wichita, Topeka, Iola, Dodge City, Kansas City, Concordia, and Saint Joseph, in regions east of the Dust Bowl proper. Alternatively, Kansas was host to more than 152 cooperative observers, with 40 of these observing stations in the western region. Similarly, in 1936 the Weather Bureau ran 15 official stations in Texas, while the cooperative force accounted for an additional 237 stations around the state.

Cooperative and official data stations were gathering fundamentally different types of meteorological data. Cooperative data was used for climatology, and so even though the bureau had cooperative stations located in the direct line of the storms, their data was never intended to be utilized for storm forecasting or reporting. Established in 1891 as a "volunteer" service when the newly inaugurated Weather Bureau took over weather data collection efforts from the Signal Corps, cooperative data collectors had the purpose of "establishing the climatology of the United States."[72] They produced a "mass of climatological data . . . literally millions of records" as part of the climatological service of the Weather Bureau.[73] These data formed climate archives for the bureau, and sometimes data for weekly weather and crop bulletins. Importantly, cooperative stations took data readings once a day, and data was sent monthly by postal mail—a parallel settler infrastructure—to Weather Bureau regional offices. In contrast, official stations produced hourly records with extensive instrument sets and sent their materials daily via telegraph to regional and national stations.[74] So, not only were official stations not located in the primary region of crisis, but cooperative stations were not equipped with the ability to measure or transmit essential data in the ways that official stations were.

As dust enveloped horizons and epistemological landscapes, Weather Bureau employees and cooperative observers worked within the confines of existing categorical arrangements on their weather data sheets.[75] Two different modes of data capture were available through bureau-issued forms—one for cooperative observers who were gathering climatic data and the other for station officials who were gathering daily weather and forecast data—but neither modes included existing categories for measuring dust. The following examples from April 1935, a month with one

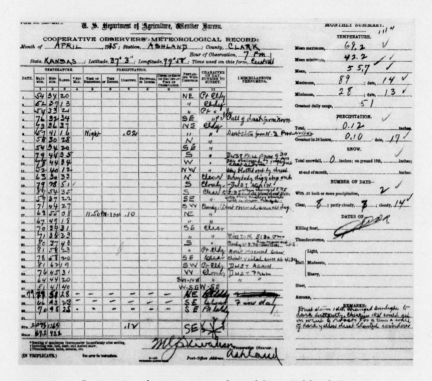

FIGURE 4.1. Cooperative observer meteorological form, Ashland, Kansas, April 1935. Courtesy of NOAA's NCEI, Center for Weather and Climate.

of the worst storms of the Dust Bowl, reveal how dust permeated several columns on the data blank, finding no permanent home within any of them. A cooperative observer data sheet from Ashland, Kansas (figure 4.1), shows how dust description gets incorporated into the "Miscellaneous Phenomena" column as well as the "Remarks" column, columns designated for anomalies and further explanation.

Between April 4 and 16, the observer has written: "Ball of dust-storm, noon," "Dust storm from N—3 p.m.-AWFUL," "sky blotted out by dust," "Everybody digging out." And from April 22 to 26: "Clouds veiled sun at 4:30," "dust again," "dust again." In the "Remarks" column, the observer wrote, "Dust storm changed sunlight to dark instantly. . . . For a time, a swell of dark yellow dust blanketed windows."

A data blank from Liberal, Kansas, from March 1935, shows how the cooperative observer struggled to decide where dust data was supposed to be entered on the sheets. Dust data oscillated between the "Remarks" column, the "precipitation" column, and the "wind" and "state of the

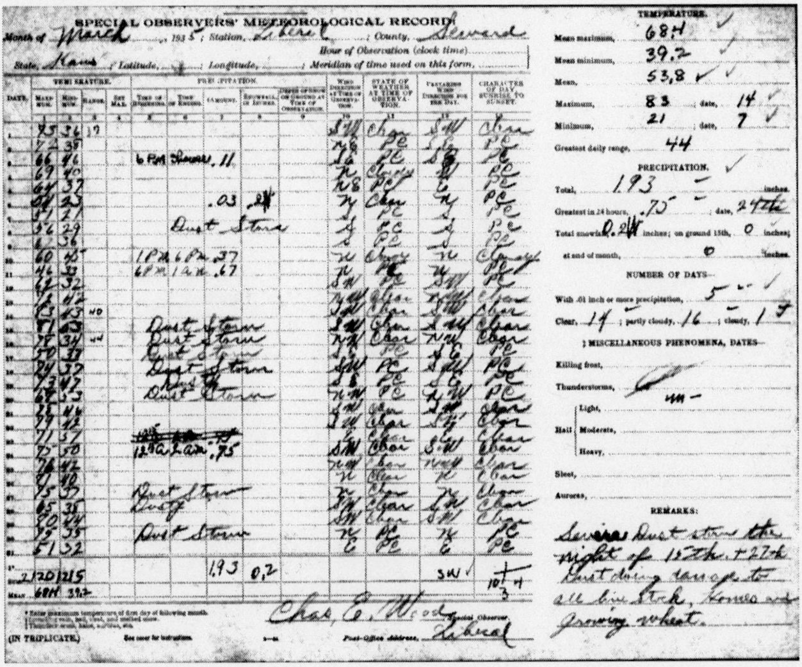

FIGURE 4.2. Cooperative observer meteorological form, Liberal, Kansas, March 1935. Courtesy of NOAA's NCEI, Center for Weather and Climate.

weather at time of observation" columns. It also oscillated between the graphical and lexical register (figure 4.2). These recordings also positioned dust within the precipitation, sun, wind, and remarks columns. Observers dealt with dust recordings by experimenting with their placement on the meteorological form, moving remarks between columns mostly designated for numerical measurement. Dust was, in fact, everywhere—dust as veil; dust as blanket; dust as swell.

The rendering of dust on meteorological forms makes clear the categorical messiness of this material. But dust was categorically messy in different ways for official station agents as opposed to cooperative observers. Official station agents used a different form for measurement, meteorological form 1001, which allowed for greater data capture. Measured items on this form included attached and unattached barometer, dry and wet thermometer, self-registering thermometer, wind velocity and direction, precipitation, and clouds (kind, direction, and amount) taken two or three times daily. Observers also measured sunshine, "character" of sunshine, wind direction, and wind velocity on an hourly basis and

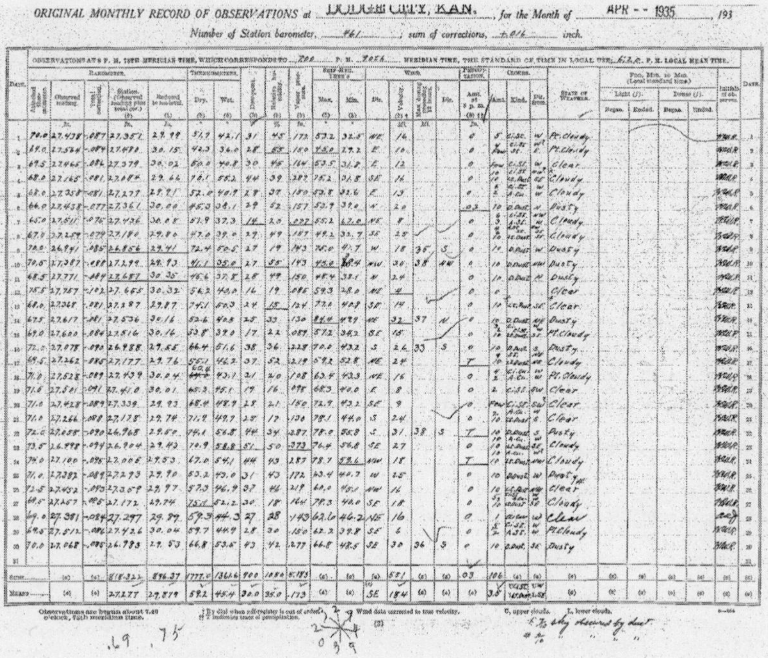
ORIGINAL MONTHLY RECORD OF OBSERVATIONS at DODGE CITY, KAN., for the Month of APR - - 1935, 193

Number of Station barometer, 461; sum of corrections, +.016 inch.

FIGURE 4.3. Weather Bureau employee meteorological form, Dodge City, Kansas, April 1935. Courtesy of NOAA's NCEI, Center for Weather and Climate.

utilized thermographs every day. Finally, observers had access to a special observation sheet, as well as a monthly meteorological notes sheet. Official observers made sense of dust differently primarily because they had access to different data collection instruments and categorical space for elaboration on their forms. At the official Weather Bureau station in Dodge City, Kansas, notations indicating dust appeared in the hourly sunshine log for the month. To render dust data in this log, the Dodge City station agent created a symbol to insert into applicable columns on the data sheet that indicate "sky obscured by dust" (figure 4.3). The symbol appears over forty times across this observer's monthly records within the columns designated for "Sunshine" (figure 4.4). The symbol is repeatable and standardizable, within the closed system of this observer's weather data records.

At official stations, the Weather Bureau asked that "special attention . . . be given to any unusual phenomena that may indicate the presence of wind-borne dust in the air" via observer reports mailed to the cen-

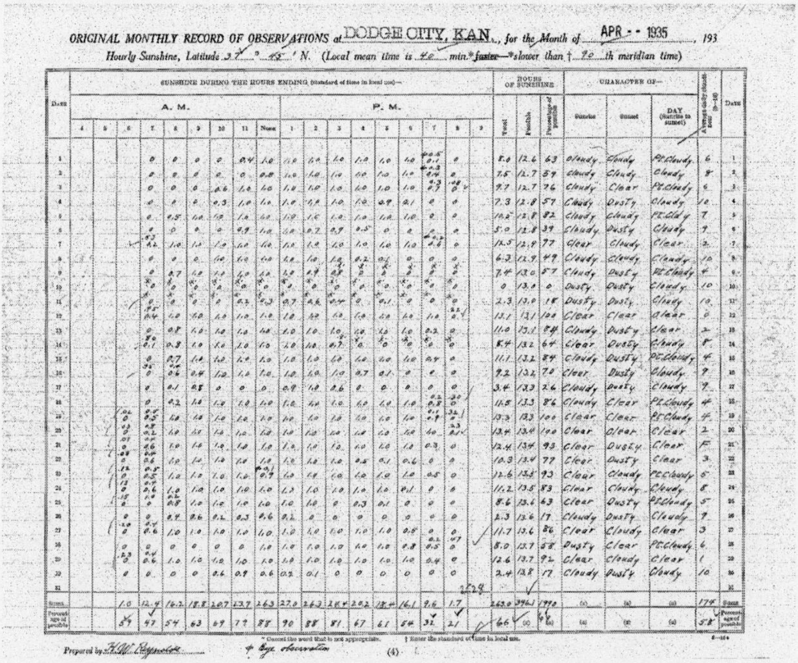
ORIGINAL MONTHLY RECORD OF OBSERVATIONS at DODGE CITY, KAN., for the Month of APR - - 1935, 193

Hourly Sunshine, Latitude 37° 45′ N. (Local mean time is 40 min. ~~faster~~—slower than 90 th meridian time)

FIGURE 4.4. Weather Bureau employee meteorological form, Dodge City, Kansas, April 1935. Courtesy of NOAA's NCEI, Center for Weather and Climate.

tral office. Station officials were asked to submit statements showing the "days with dusty conditions, the intensity as measured by the visibility, duration, special remarks, and . . . photographs."[76] Yet, they also emphasized that "if no dust was noted, no report is desired."[77] Even the Weather Bureau was unsure about the best way to gather data on these storms from land-based stations. What "unusual phenomena" would indicate the presence of wind-borne dust? Without dust-measuring instruments, employees had to rely on their senses—a radical reorientation from automatic sensors and self-recording instruments—to discern the presence of dust. Because employee records were supposed to be based on a variety of self-registering instruments, the station agent noted which observations were self-registering, automatic, or mechanical versus made through "eye observation."

Both employee and volunteer weather data records from this period reveal the extent to which dust had not only invaded the physical landscape but also confused the data landscape. Cooperative forms and form 1001 show an inability to numerically measure the dust, and a turn toward qualitative and symbolic expression as a dominant way of registering dust. Dust was an obstruction to vision, but what particular vision was it obstructing? On both forms, data collectors tried to fit the crisis of dust into an already established system of measurement. On the volunteer forms, we see observers allowing dust to impact their measurements across precipitation, general sky and weather conditions, and miscellaneous phenomena. Employee records fit dust into the columns for clouds and state of the weather. Beyond this, employee records are less elaborative of the quality, shape, and size of the dust in columns meant for quantitative—designated for numerical or linguistic shorthand—readings. So, rather than registering "veils" or "balls" of dust as the volunteer observers do, employees tend to stick with descriptors of "light dust" or "dusty." Finally, we see more expressive descriptions of dust across volunteer observer sheets and more systematic innovation across employee sheets.

Dust Bowl data collection reveals what happens when measurement regimes encounter events and materials (dust storms and dust) that they were not designed for. These forms were never meant to capture dust storms. Yet, in these attempts to measure, we see how dust data fell to the edges of the form—the corners of precipitation and sunshine record boxes, the margins of miscellaneous phenomena. Studying data's edge here allows us to ask how existing modes of data capture deal with new stimuli. It also allows us to inquire into how failures of land-based data capture can signal toward another environmental knowledge–making world. The difficulty that volunteers and employees encountered on the ground when they attempted to measure and register dust data suggested a fundamental disjuncture between historic forms of colonial scientific capture and complex earth systems.

From data capture to sodbusting, settlers narrativized environments as capital markets and treated them as objects to be studied, rather than subjects to be understood in relation.[78] Settler environmental consciousness upheld a notion of bounded, object-oriented weather, exemplified in the graphical layout of meteorological forms and the discourse of clean and dirty—good and bad—data. Bad data was the culmination of a century's worth of data collection practices that were rooted in meteorological science's embeddedness in settler colonial worldviews. Bad data was data on the edge—data that a form cannot capture. Bad data is dirty data, which is data on the brink.

The Dust Bowl was not the only crisis that influenced meteorology in the mid-twentieth century. World War II brought more money and interest to the field of atmospheric sciences, as professionally trained meteorologists assisted in war and postwar science, and the theoretical development of meteorology—specifically the exploration of upper-air conditions—was spurred by both world wars.[79] Military aviation as well as providing weather services to military forces increased in 1917, when the United States officially entered World War I. Between 1930 and 1939, Weather Bureau appropriations were devoted to aiding aviation.[80] The relationship between aviation and meteorology grew when, on January 28, 1937, the Air Weather Service—located within the Air Corps and the larger Department of War—was created through an order from the secretary of war. The Air Weather Service inherited 40 weather stations, 22 weather officers, and 180 enlisted men from the Signal Corps. An additional 180 enlisted men were transferred to the Air Weather Service in July.[81] The Weather Bureau assisted the War Department throughout World War II. Some bureau officials were assigned to military locations abroad; they jointly studied stratus clouds in California with hopes of applying this information to the conditions of military flights and worked together to leverage current weather and forecast predictions for use in military strategy.[82] Although oriented toward different goals—war as opposed to agriculture—both the War Department and the Weather Bureau recognized the necessity of aerological observations.

Both Weather Bureau meteorologists and academic meteorologists desired to explore and support research in air mass analysis and long-term forecasting prior to and throughout the Dust Bowl period. In 1933, the secretary of agriculture asked the Science Advisory Board to submit recommendations for improvement of the Weather Bureau. The Science Advisory Board released its report in 1934, recommending that the Weather Bureau pay more attention to data collection from the air through adoption of the "air-mass analysis method of weather forecasting and the consolidation of the system of recording and reporting meteorological data."[83] In response, the Weather Bureau established twenty aerological, upper-air stations across the country, working with the Air Corps of the army and the Bureau of Aeronautics of the navy to establish these stations. Additionally, in 1927, the Committee on Aeronautical Meteorology was formed with the task of advancing "the art and science of meteorology in aviation."[84] The Weather Bureau as a whole continued to face stagnation and "dwindling appropriations," and so cooperation between the Air Corps, the Bureau of Aeronautics, and academic meteorologists was key for vertical data collection and analysis during dust activity and in the decades after.[85]

In 1936, the *Bulletin of the American Meteorological Society* published a paper titled "Dust Storms over the Great Plains: Their Causes and Forecasting."[86] G. R. Parkinson—meteorologist for Transcontinental and Western Air in Kansas City—utilized air-mass analysis to analyze dust storms. Parkinson, after having been stationed at Amarillo, Texas, during the 1930 to 1931 dust storms, was again stationed in Kansas City from 1932 to 1935, where he utilized meteorological data from Transcontinental and Western Air operating routes—routes that flew "over the heart of the dust storm country or so called 'dust bowl' from Kansas City to Albuquerque."[87] Parkinson used free-air sounding and aerological observation data to study the characteristic properties of air masses that caused dust storms. Parkinson's argument pointed to the ways that conditions on the ground and in the upper air—such as crop cover, wind, and cold fronts—were essential for determining dust storm logic. His work also underscored that these interlaced conditions could only be adequately assessed and understood from the air, something that interested both theoretical and applied meteorologists. Air-mass analysis—based in horizontal and vertical data, as well as theoretical models—provided comprehensive portraits of storms.

On long-range forecasting, the Bankhead Jones Act of 1935 aided in providing financial support "to conduct research into laws and principles underlying basic problems of agriculture in its broadest aspects." It allowed for the emergence of long-range forecasting at the Weather Bureau through government-university partnership. Dr. Charles F. Sarle, the principal economist with the Bureau of Agricultural Economics of the Department of Agriculture, worked with Professor Carl-Gustaf Rossby, head of the Department of Meteorology at the Massachusetts Institute of Technology (MIT), to review and evaluate "previous efforts in long-range weather forecasting, the organization and analysis of global data sources and the preparation of experimental extended-range forecasts."[88] By 1938, MIT was running forecast experiments that produced five-day forecasts with "mean Northern Hemisphere circulation and surface temperature and precipitation" and "daily prognostic charts of 7 days."[89]

While the Weather Bureau and the Air Weather Service had been able to collect massive amounts of weather data over the course of the early century and make some progress toward long-range forecasting in cooperation with MIT, they were not ultimately able to use "a numerical approach to solving the non-linear equations defining atmospheric movements."[90] Applied meteorology and data collection could not address these questions; rather, a theoretical approach was necessary. This approach would unfold throughout the 1940s and 1950s as operational numerical weather prediction grew through the Meteorology Project at the Institute for Advanced Study at Princeton.[91] As numerical weather prediction progressed through interagency cooperation and the work of academic meteorologists, the next major move in weather data collection and visualization—aimed at developing a "truly global" imaging technology that removed the problem of bad, or incomplete, data from the equation (or so it seemed)—was also in the works.

In the Dust Bowl period, a move to the air allowed meteorologists to think vertically about data collection and meteorological science, connecting the ways in which dust storms were bound not to a single category on the data sheet but rather to every single one. Yet, meteorologists, and the larger public that meteorology served, still struggled to think relationally. The problem remained that settlers could not see—or did not care to see—themselves in the stirring dust. While federal legislation through the New Deal attempted to solve some of the practice-based issues that caused the Dust Bowl, the environmental consciousness—rooted in environmental domination and atomism—that caused the dust storms remained active well beyond the 1930s, spurring additional injustices in the century to come whereby "settlers' lack of long-term knowledge of the environments they inhabit even years after settlement" "have transformed the ecological conditions in ways that are not sustainable for settlers, Indigenous peoples, or anyone else."[92] Data capture, weather crisis, and settler colonialism converged in the Dust Bowl period when land-based data collectors treated dust discretely—as a singular object rather than a multidimensional subject to be understood in relation. This converging, as the final chapter of this book will show, is still present today. The weather mixes everything up.

Across the 1960s, weather data began to shape popular visual media. Data soon became a visual media with the onset of meteorological satellites. Meteorological satellites were directly tied to the visualization needs of the Dust Bowl period and the rise of military aviation, which required upper-atmospheric data collection. These historical forces fostered a historical transition between graphical-numerical data and newly photographic data. This transition came into full swing with the launching of the first meteorological satellite, TIROS. From the beginning, TIROS—the United States' first low-earth satellite—was imagined as "a truly global means for observing a truly global phenomenon": weather.[93] What Eddy had begun in his backyard—an experimental venture that linked weather data collection and the photographic visualization of terrestrial space—reached a new height in the meteorological satellite movement of the 1960s. While TIROS was not the first weather technology to gather data from the air (as Eddy exemplifies), it was the first supposedly "truly global" means for imaging weather and collecting weather data. It was also, importantly, the first weather technology capable of producing global products for a mass-media audience. The images from TIROS, much like Eddy's, made their rounds through various publications and media outlets; they spurred a national environmental consciousness that sought to quantify the whole of the biosphere.[94]

5

UGLY DATA IN THE AGE OF SATELLITES AND EXTREME WEATHER

Dr. Hans K. Ziegler was not originally slated to give the luncheon address at the July 1969 Fifth Annual Technical Exchange Conference of the Air Force Academy held in Colorado Springs. Major General William B. Latta, the commanding general of the US Army Electronics Command at Fort Monmouth, was supposed to do so, but at the last minute he could not make it. Rather than read Latta's prepared speech, Ziegler wrote his own, one that focused on what he understood to be one the army's "greatest meteorological achievements of our time"—the Television InfraRed Observation Satellite (TIROS). Ziegler knew how important TIROS had been to the US Army because he had acted as the chief scientist of the laboratory that brought the satellite from a think tank reconnaissance concept to a space-ready weather data machine.

In his speech, Ziegler—deputy for science and chief scientist for the US Army Electronics Command—elaborated on the role his team played in developing the "meteorological aspect" of TIROS, a role bequeathed to him and his team from Major General John Medaris, head of the Army

Ballistic Missile Agency. As Ziegler elaborated in his speech, the plan for a target acquisition system—what would later become the world's first meteorological satellite—required knowledge of targets in "the grid of the earth," and such "accurate information could only be provided by a satellite-based system."[1] Ziegler discussed the shortcomings of TIROS as a pure target acquisition system, as well as "the Army's important present and future needs in weather information and meteorological and environmental data" about which he was "deeply concerned." The needs Ziegler listed were immense, ranging from "improved ballistic meteorological support" to "terrain weather support to plot our moves on the surface of the earth" and "climatological information for overall Army-wide planning and utilization."[2]

Ziegler's speech is important because it narrates the US Army's relationship to meteorological data collection as both a recent past and a then-urgent future. Additionally, it highlights the historic interdependence of the Weather Bureau, the National Aeronautics and Space Administration (NASA), and the military sectors of the nation-state (from the army to the Signal Corps) in seeking, collecting, and utilizing meteorological satellite data. TIROS was itself a product of the century-long cooperative relationship between the War Department and the Weather Bureau. As outlined in this book and elsewhere, the Signal Corps and the US Army collaborated on many occasions to utilize weather data for the purposes of nation-building, settlement, land dispossession, and military meteorology at home and abroad. In chapter 1, I addressed how settlement across the eastern region of the nation necessitated weather data collection and outlined how some of the earliest leaders in data collection also acted as surveyors and land speculators. As chapters 2 and 3 address, military forts across the settler states of the nineteenth century were also data collection outposts. In the nineteenth century, meteorological data collection was necessitated by, and fed, settler colonial aims.

In the twentieth century, however, the relationship between meteorological data collection and settler colonialism played out through the acceleration of partnerships between nation-state agencies like the War Department, the Central Intelligence Agency (CIA), the Department of Defense, the Weather Bureau, NASA, NOAA, and others. Through these partnerships, meteorological science played a crucial role in foreign and domestic statecraft, from war to resource extraction. As we have already seen, the War Department and the Weather Bureau have a long history of collaboration. As explored in chapters 2 and 3, the War Department inherited aspects of the Smithsonian Meteorological Project in 1870, prior to the establishment of the Weather Bureau in 1891.[3] In the early part of the twentieth century, the Signal Corps gifted the Weather Bureau pilot balloon equipment and five kite centers, which had been used during World War I.[4] During that war, the Weather Bureau and the Signal Corps worked together on military meteorology, including "battlefield climatology." The US Army and US Navy "trained approximately 8,000 weather officers" during World War II, in addition to establishing the Air Weather Service.[5] The acceleration of interagency partnership in the middle to late twentieth century, then, comes as no surprise.

The TIROS program emerged from the wake of these historic relationships and resource sharing. Just as previous partnerships and technologies had paved the way for TIROS, TIROS laid the groundwork for satellites to come, most significantly the rest of the TIROS fleet (TIROS 2 through 10, and TIROS-N) and the Geostationary Operational Environmental Satellites (GOES 1 through 17). This chapter explores this era of satellite meteorology and its relation to shifting national environmental consciousnesses by investigating the weather data cultures US settler meteorology espoused. The nation-state's turn to "global" data in the 1960s—where this chapter begins—extended two historic settler ideologies. First, it extended the idea of knowing through quantification, where quantifying the weather was bound to satellite imaging technology. Second, it extended the now century-and-a-half-old alliance between meteorological data collection and the consolidation of the nation-state's power. Across the two centuries this book considers, meteorological data collection required constant interface with formal branches of the nation-state, from the Department of War in the nineteenth century to the CIA and the Department of Defense in the twentieth century. Meteorological data collection was, and is, a form of power consolidation, and in the twentieth century this form of power was utilized both at home and abroad. Additionally, while NASA operated based on free and open nation-based science, the Department of Defense did not, and the department used that same free and open science for military reconnaissance and strategic warfare. Through data capture technologies such as meteorological, and defense-meteorological, satellites, environmental data has had specific uses and applications under petrocapitalism. These uses have extended the power of the nation-state while also exacerbating resource extraction and climate catastrophe.

This chapter chronicles the role of satellite meteorology as it sits at the end of the two-century drama of the US settler state's utilization of weather data and weather data technology toward world-building ends. From TIROS and NIMBUS to Landsat and GOES, satellite meteorology has dominated weather data collection, collation, and visualization from the mid-twentieth to the early twenty-first century. The first section of this chapter explores the development of the TIROS satellite as a military reconnaissance meteorological project, and specifically the ways in which meteorological data served the settler colonial ends of international warfare. The next section uncovers how collaboration and satellite data resource sharing between NASA and the oil and gas industry through Landsat—a satellite made possible because of its predecessors TIROS and NIMBUS—facilitated resource extraction across the US settler states. The chapter concludes by considering the late twentieth-century turn toward mass media weather data visualizations, analyzing the meteorological satellite movement of the 1960s, contemporary GOES 14 experimental satellite technology, and visual satellite data and media during storm crises. I explore how in the early twenty-first century, satellite views—and the data they were built on—became quotidian features of environmental crises, such as Hurricane Sandy. Yet, despite the complicated process of data gathering and transmission that is involved in meteorological satellite sensing and imaging, satellite media often pass as uninterpreted weather reality for popular audiences, offering visual and textual narratives of weather crises that abstract the US settler state and its inhabitants from environmental disaster and the responsibilities such relations bear.

Meteorological satellites and their data products restructured the vision inhabitants of the nation-state had of themselves, the weather that surrounded them, and the nation they belonged to. The meteorological data culture of the 1960s on prioritized twenty-four-hour data streams and constant access to satellite data meteorological products. Yet, the visual products that circulated as satellite data encouraged environmental narratives that removed the situated and haptic body from weather and storm crises. In many ways, the conceptualization of weather data had reached its logical conclusion now that it could be formulated as separate from the settler body. Importantly, satellite-based weather products and the data they are built on have had ontological ramifications for environmental imaginaries and especially narratives of climate and climate crisis in the contemporary US settler states. Visual satellite data present a disembodied, abstracted version of "the human" with respect to weather in general and weather crises in particular, whereby historical relationships of domination, theft, and abuse are absent from the storytelling of environmental chaos. This dissociation is both a feature of settler weather data culture and a crisis that threatens the capacity of settler science to reimagine itself beyond its current system thresholds.

MILITARY METEOROLOGY AND THE EVOLUTION OF THE TIROS SATELLITE

In 1946, the US Army Air Force initiated a research and development project with Douglas Aircraft Company. This project produced a new research entity, RAND (Research and Development, and later known as the RAND Corporation), which was tasked with exploring the creation of a "world circulating space ship." Staff and consultants met for three weeks in April 1946 and, at the end of this three-week period, published their first report: *Preliminary Design of a World-Circulating Space Ship.*[6] In this report, they noted that while "the present study was centered around a vehicle to be used in obtaining much desired scientific information," the vehicle will also "undoubtably prove to be of great military value."[7] From the beginning, national satellite meteorology and military strategy traveled hand in hand. Consultants for RAND valued the opportunity for surveillance that a world-circulating space ship would provide. The consultant Lyman Spitzer Jr., a professor of astronomy at Yale, speculated on the "important property" of a satellite to provide a "platform from which a very wide expanse of the earth can be viewed," noting the benefit of detecting moving ships.[8] The RAND staff also proposed the use of "television equipment," producing an "observation and reconnaissance tool without parallel."[9]

In 1949, RAND sponsored a conference centered on the "utility of space satellites." Though RAND's first study outlined the scientific benefits of a "world-circulating space ship" with only a brief mention of military use, by the time of its 1951 report, "Inquiry into the Feasibility of Weather Reconnaissance from a Satellite Vehicle, R-218," the intellectual groundwork had been laid to consider the military purposes of a meteorological satellite. The report read: "It is assumed that, in the event of an armed conflict, aerial weather reconnaissance over enemy territory, similar to that obtained in World War II, will be extremely difficult if not impossible. An alternative method of obtaining this information, however, is thought to lie in the use of the proposed satellite vehicle."[10] If, in 1946, RAND upheld that the primary purpose of a world-circulating spaceship was to gather scientific "information," or data, then it was clear by 1951 that one of the main purposes of RAND's conceptual work was indeed to aid the military through weather reconnaissance.

Under a classified program called Project Feed Back, authorized by the US Air Force, RAND not only theorized the meteorological reconnaissance satellite but also engaged the Radio Corporation of America (RCA) in contract studies regarding the use of televisual techniques in 1949.[11] Through funding from the air force, RCA assigned eighteen professional scientists and engineers to research the use of television for "reconnaissance purposes."[12] Importantly, Project Feed Back was not just about meteorological satellite reconnaissance. It was also about handling, transmitting, and evaluating data. Aptly titled, the project aimed to create an "intelligence data chain," whereby data would "feed back" between air and ground stations across the globe, including "perhaps five, near the periphery of Russia."[13] Under the Project Feed Back proposal, the satellite was meant to see a range of things, only a few of which were actually meteorological in nature—airfields and the presence of large planes; industrial concentrations; large plants; harbors, docks, and large ships; transportation, power, and communication networks, including switching yards, bridges, canals, power lines; urban areas; large military installations, including camps and explosive storage; and, finally, cloud patterns and structure.[14] The intelligence data chain provided by this reconnaissance satellite was predicted to be able to provide data for "disruption" missions (whereby a satellite provides reconnaissance for an attack mission) and "bomb-damage" assessments (whereby a satellite assesses bomb damage in times of war). The feedback system included an "intelligence center" for "data reduction" and "storage," where "selectors, interpreters, and analysts would work together" to process new data.[15] Much like data collection earlier in the century, Project Feed Back assumed use of and access to physical space—land—through satellite construction and resource allocation, storage, and launch alongside the physical space necessary for data processing itself.

Despite RAND's continued study of the practical application of a meteorological reconnaissance satellite throughout the early 1950s, the air force did not pursue the building out of such a concept.[16] However, in 1957, William W. Kellogg—one of the key researchers of Project Feed Back—was asked to head an advisory committee within a new government agency, the Advanced Research Projects Agency (ARPA), housed within the Department of Defense. One of the goals of Kellogg's advisory committee within ARPA was to incorporate the requirements of "Inquiry into the Feasibility of Weather Reconnaissance from a Satellite Vehicle, R-218" into a satellite project. This was project TIROS.[17] Through ARPA, RCA's Astro-Electronics Division designed and built TIROS under the direction of the Army Signal Corps Research and Development Laboratories in Fort Monmouth, New Jersey. However, by 1958, the goal of the project had shifted from reconnaissance to meteorology. Infrared sensors were added to the project design, and the project was formally titled the Television InfraRed Observation Satellite.[18] The Army Signal Corps Research and Development Laboratories directed the construction of the weather satellite. After the satellite's 1959 transfer to NASA—an association, Neil Maher argues, that was tied to the military, despite its founding by President Eisenhower as a civilian agency—and its launch in 1960, the Army Signal Corps Research Development Laboratories also directed one of three command and data acquisition stations for incoming TIROS 1 data, or what had been imagined as one node within the RAND 1951 feedback system's "intelligence center."[19]

The time, money, and think tank theorizing that went into RAND's concept for a weather reconnaissance satellite vehicle—outlined in Report 218 and funded by the US Army Air Force in 1946—eventually resulted in the satellite that the Army Signal Corps was tasked with building and that Ziegler himself oversaw as chief of the Fort Monmouth, New Jersey, laboratory. This is the achievement he would celebrate in 1969 at the Fifth Annual Technical Exchange Conference of the Air Force Academy. And it is this achievement that would shape public and private access to the next generation of weather data. As a product of ARPA, TIROS originally carried a confidential classification. But given NASA's mission toward free exchange of environmental information, it did not wish to classify TIROS or the meteorological data TIROS produced. NASA planned to release data to the general public, some of which might include imaging of classified locations. This created a problem for US intelligence and security, according to the US Intelligence Board, in that TIROS made clear that photographic reconnaissance from space was indeed possible. In 1960, the Soviet magazine *International Affairs* speculated that TIROS, although publicly announced as a cloud-cover photographing satellite, was being used by "American military leaders" for "espionage."[20] The optical capabilities of TIROS had been downgraded from its original design of a 125 kilocycle video bandwidth to a 62.5 kilocycle bandwidth when the project was reoriented from a reconnaissance satellite to a meteorological one in

1958.[21] In its reclassification, TIROS was rendered incapable of the kind of photographic reconnaissance it was initially designed for and that the Soviets feared.

However, TIROS was not the only meteorological satellite in the sky in the early 1960s, and there were indeed satellites capable of such data collection and imaging. Project 417—also called Program 35 and Program 698BH—was a highly classified meteorological reconnaissance satellite project directed by Lieutenant Colonel Thomas Haig of the Defense Meteorological Satellite Program (DMSP), which was housed in the National Reconnaissance Office (NRO). Project 417 utilized program design and hardware from TIROS activities, and project officials negotiated first with NASA and then with NASA's booster producers for components of the airframe system.[22] After several failed launches, Project 417 placed a cloud-cover satellite in a sun-synchronous orbit in 1962 that allowed for daily weather coverage of the Soviet Union. The satellite also provided weather pictures of the Caribbean, which were utilized during the Cuban Missile Crisis.[23] Launch and operation were handled by the air force.

The project TIROS was a successful meteorological satellite whose reconnaissance capabilities had been stripped to allow it to operate as a public-facing meteorological data collection project. The TIROS technology also allowed for other reconnaissance systems within the DMSP. With public support behind the TIROS in the sky, the Department of Defense—and more specifically the NRO—could focus its energy on utilizing TIROS engineering plans and hardware to craft the meteorological defense satellites RAND had outlined in its series of reports from 1952 on. In July 1959—several months before TIROS's launch—the CIA released an internal memo regarding the intelligence aspects of Project TIROS. In it, the agency speculated that NASA's release of the relatively "poor quality" TIROS pictures would "convey a false impression" as to the imaging possibility of a camera-carrying satellite; TIROS was used by the intelligence community to cover the very real technological capabilities of DMSP.[24] Eisenhower, too, understood that TIROS would "shield" the more covert activities of the DMSP—including Corona, Samos, and Project 417—and would also provide a narrative of "peaceful" space exploration.[25]

Under NASA's control, TIROS exemplified the free and open exchange of meteorological data. Yet, the NRO and CIA's emerging reconnaissance data economy—based in the same satellite technology as TIROS—was diametrically opposed. The interaction and overlap between TIROS and DMSP raised challenging questions about the public and private nature of meteorological data. Whether it was public or private, one thing was abundantly clear—photographic and televisual data was king.[26] In *Beautiful Data*, Orit Halpern documents how vision became a "universal language." Halpern writes that, following World War II, there was a "faith that vision, above all, could reconcile the humanities, sciences, and arts; providing a universal language not only within the university but for what was increasingly understood to be a global and interconnected planet."[27] This was a turn, according to Halpern, not only toward vision but, more important, toward the objectivity and sovereignty of vision as a primary way of knowing. This newfound belief in vision was closely tied to a belief in the objectivity of data. This emphasis on vision, on human sight, was aligned with the promise of *seeing* "vast data fields," a turn away from organizing a "perfect record" toward the "management and organization of patterns and the construction of dynamic structures" that promised not only to be visual but also to reveal knowledge.[28]

The power of visual knowledge in the military-meteorological sector rested on accumulating and processing visual data. Whether from civilian or military satellites, all data had to pass from the satellite to receiving stations and then to a central command hub—the "intelligence center" imagined in RAND's 1951 report. The military developed a satellite command center—the Air Force Satellite Control Facility in Sunnyvale, California—that was separate from NASA's command and control network.[29] Defense infrastructure and intelligence supported data transmission and collection, from reconnaissance satellites to meteorological satellites. The Kaena Point, Hawaii, tracking station was a military installation of the US Air Force established in 1959 to support the classified Corona satellite.[30] This station obtained data from all reconnaissance-based satellites and also acted as one of the two primary remote tracking stations for the TIROS weather satellite and its predecessors. The station in Fort Monmouth, New Jersey, played a similar role: it obtained data from both types of satellites. While control facilities were separate entities, the tracking stations were not. There was considerable overlap between the data collection and tracking stations that supported transmission of both reconnaissance (NRO) and civilian (NASA) satellite data.

The turn to meteorological satellites as devices for data collection—and especially visual data—allowed for, in the words of Lisa Parks, "technologized vision" where the act of witnessing "demilitarize[s] military perspectives, . . . open[ing] the satellite image (and other forms of image data and intelligence) to a range of critical practices and uses." As Parks suggests, this demilitarizing travels hand in hand with the satellite image's "aesthetics of remoteness and abstraction." Yet, whatever "lies in the field of vision" produced by the satellite "implicates the state."[31] Meteorological satellites, and the forms of image data and intelligence they create, were demilitarized through the framework of NASA. Despite their historical and contemporary ties to military reconnaissance, defense, and natural resource extraction (investigated in the next section of this chapter), meteorological satellite data appear not through the military apparatus of the satellites' creation but through an "aesthetics of remoteness and abstraction." This has allowed and emboldened a weather data culture in which nation-state technologies for reconnaissance, resource extraction, geophysical domination, and meteorological data collection have become eternal bedfellows, though they rarely announce themselves as such.

With two television cameras mounted on the inside of the craft, TIROS successfully launched into space on April 1, 1960. Its photographs were stored on a magnetic tape recorder and sent back to primary ground stations in Fort Monmouth, New Jersey, and Kaena Point, Hawaii, when the satellite was in range. The project TIROS was cheered as a gift of technology by NASA scientists and meteorologists and was articulated as "a truly global means for observing a truly global phenomenon," the weather.[32] The "truly global" aspirations of TIROS were dependent on the reconnaissance theorizing that brought it into existence. Though TIROS was a "global" satellite for the public, it was indeed a machine that made possible particular nation-building tasks—from reconnaissance to natural resource extraction—for the nation it served.

During its 1,302 orbits around earth, TIROS captured twenty-three thousand "pictures." The data collected was a pure output of vision, of "television cloud photography." The satellite's images made their rounds through various publications and media outlets. On May 1, 1960, the *New York Times* published the TIROS satellite pictures and a three-page article on TIROS's image and data technology.[33] The article observes that the weather satellite "promises to come to the meteorologists' rescue" by offering instantaneous "information" and "data" about "whole sectors of the upper atmosphere." It emphasizes the new fields of vision provided by the satellite pictures and all the data that comes along with that vision. That same year, the *Monthly Weather Review* published a seventeen-page image and text report on the TIROS pictures.[34] Throughout the report, TIROS's images are described as palimpsests of weather data and not as aesthetic objects: "cloud pictures" that "portray the details of the cloud structure around portions of [a] cut-off cyclone in the eastern Pacific in strikingly clear fashion."[35] The emphasis is on the clarity of cloud detail as seen from space, something that before TIROS was unseen by meteorologists and the general public; the emphasis is especially strong in the *Monthly Weather Review*, where the images are superimposed with latitude-longitude grids and wind-contour data.[36]

The project TIROS reveals an unresolved discourse around pure visions of data, a discourse that would follow meteorological data culture and its consumers from TIROS to GOES across the second half of the twentieth century and into the twenty-first. Pictures produced by TIROS showed the ways that satellite vision and data is never seamless but is always being recontextualized—gridded and calibrated with additional data so that it can be "read." The satellite was able to get clearer and wider pictures of earth, furthering the science of cloud and storm analysis as well as prediction. But it could only further these scientific endeavors when the pictures were viewed as messy media that needed to be calibrated and cleaned, gridded and superimposed, in order to reveal weather truths. While the calibration and cleaning of data were part of public discourse, what was less clear in the public domain were the ways that TIROS and its soon-to-be successors were implicated in a data economy that continued to harness the biosphere for settler colonial presents and futures through interdependences between satellite meteorological data collection, government reconnaissance, and resource extraction in the United States.

DATA RICH AS OIL

TIROS was the first operational weather satellite, and its broad success shifted the course of meteorological remote sensing. In 1964, four years after the successful launch of TIROS, NASA launched NIMBUS, a weather satellite concept developed by NASA and the Weather Bureau. The NIMBUS satellite was originally designed as "the future operational weather satellite," yet throughout its early life (NIMBUS 1 through 4), the satellite played an experimental role in that it tested new data collection and acquisition techniques. Each satellite included an automatic picture-taking (APT) vidicon capability, a data collection technology that had been first implemented on TIROS 8 in 1963. Built by RCA, the APT camera "utilized a very-slow-scan vidicon" as opposed to the wide-angle and narrow-angle half-inch TV cameras that were utilized on TIROS 1 through 7. The increased 2 kHz bandwidth of the APT system allowed

TIROS 8 to "transmit direct, real-time television pictures to a series of relatively inexpensive APT ground stations located around the world."[37] NIMBUS altered the data collection and receiving landscape, as it generated a data collection and readout technique that increased data access worldwide.

Between 1964 and 1972, the NIMBUS 1 through 4 satellites were launched into sun-synchronous, near-polar orbit, with each of them playing an important experimental role in the development of resource data acquisition. NIMBUS was "concerned primarily with providing atmospheric data for improved weather forecasting," but the addition of experimental sensing devices allowed NIMBUS data to be used in fields such as oceanography, hydrology, geology, geomorphology, cartography, and agriculture.[38] NIMBUS bridged the divide between sensing meteorological data (specifically the remote sensing of the upper atmosphere) and sensing of earth "resource" data; NIMBUS was an outgrowth of TIROS, yet also the vehicle that paved the way for Landsat—formerly called the Earth Resources Technology Satellite Program—NASA's first earth-observing satellite with the capability to monitor resources, from crops to shale.

Landsat was an adaptation of the NIMBUS satellite.[39] Added to the basic NIMBUS bus were "a wideband recording and transmission system, the RBV (return beam vidicon) and MSS (multispectral scanner) sensors."[40] NASA's earth resource monitoring satellites were fundamentally different from both TIROS and NIMBUS yet were also deeply indebted to the existing data collection technology of both meteorological satellites. The RBV sensor was still televisual technology—as was used for TIROS and NIMBUS—that used a shutter to snap a photo. This allowed for exposure to a light-sensitive plate. The RBV was developed by RCA—the company responsible for both TIROS's and NIMBUS's vidicons—and was based off earlier vidicon cameras.[41] Additionally, the RBV was capable of a resolution of 4,500 to 6,000 lines per picture, whereas TIROS 1 was mounted with two cameras that used 500 scan lines.[42] NIMBUS 1 through 4 had experimented with a variety of spectral channels and regions: NIMBUS 4, for example, experimented with more than forty-three spectral channels and captured visible, infrared, far infrared, and ultraviolet spectral regions.[43] While the usefulness of a variety of spectral channels and regions was well established by NIMBUS and TIROS, Landsat's MSS was a fundamentally new technology that depended on a mirror reflecting light onto "a small array of photoelectric detectors, which produced an electric signal that depended on the intensity of the light reflected by the mirror from the scene on to the detector."[44] NASA also developed and incorporated a scanner for color pictures, which utilized a prism.

Although Landsat faced many hurdles in terms of government funding and interagency cooperation regarding concept and design, the first Landsat satellite was successfully launched in 1972.[45] As Maher shows, "Landsat's colorful maps quickly became critical tools for analyzing the state of earth's natural resources."[46] After its launch, both NASA and private industry began to imagine the potential use of Landsat data for the extraction industry. When NASA "invited twenty oil companies to send teams to hear about the investigations NASA had sponsored," both Conoco and Chevron expressed interest in collaborating on this research project.[47] If NASA could drum up interest in Landsat data from the oil industry, the oil companies would have a new subset of data users and would ensure "quick growth of sales and profits," thereby helping to "justify the continuation of the Landsat project."[48] Collaboration was a challenge because petroleum exploration was proprietary: the "extractive industry kept use of Landsat data secret" so as to "protect both new methods of analysis and information about what regions were being studied." The companies also did not want to work in an official capacity with NASA because the research results "would have to be made public."[49]

Despite these challenges, private industry formed the Geostat Committee on December 2, 1976. Membership was restricted to companies—a group of oil, gas, and mineral company representatives including (but not limited to) Exxon, Gulf, Chevron, Mobil, Shell, and Texas Gulf Oil—with an expressed industry interest in remote sensing application for oil, gas, and mineral extraction across the US settler states.[50] The primary objective of the committee was "the furtherance of geologically dedicated satellite systems for stimulating and improving resource exploration and engineering capabilities."[51]

The committee's objective, and its desire for access to raw data from NASA space data acquisition, ultimately led to the Joint NASA/Geostat Test Case Project, a research collaboration between NASA and the Geostat Committee in which the utility of remote sensing technology for "geological mapping in general and resource exploration in particular" was tested.[52] Remote sensing data were acquired over eight test sites "containing known deposits of copper, uranium, and oil and gas" by using laboratory spectrometers, ground-based radiometers, airborne scanners, orbital sensors, and multispectral imaging techniques.[53] The Geostat Committee would not "undertake or underwrite hardware developments," nor would it "monetarily contribute to satellite operations."[54] For this reason, the committee was entirely dependent on NASA satellite technology. Key to the joint project's study of these test sites was remote sensing data from NASA's Landsat satellite.

The Joint NASA/Geostat Test Case Project was initiated in 1977, six years after the first Landsat satellite was launched. The project used Landsat data, as well as a variety of experimental sensing systems, to evaluate the use of experimental remote sensing in resource exploration and extraction. It utilized Landsat's MSS data in digital and photographic form, Skylab photography, aperture radar images acquired by Seasat spacecraft, and two experimental sensors developed by NASA—the Modular Multispectral Scanner, airborne sensors and radiometers, and the NS-001 Thematic Mapper.[55] Test sites were chosen based on existing knowledge of deposits of copper, uranium, oil, and gas in order to test scanner and remote sensing capabilities. The project utilized NASA's Jet Propulsion Laboratory to process and enhance images, and then share with the investigative team, which was composed of geologists from Jet Propulsion Laboratory and Geostat Committee members.

Landsat's MSS and the NS-001 scanner proved fruitful for petroleum test sites. While remote sensing techniques had been previously used for structural mapping, the test project aimed to use these techniques to detect surface manifestations of hydrocarbon seepage.[56] The project successfully demonstrated the efficacy of multispectral images for the detection of hydrocarbon seepage at the Patrick Draw test site in Rock Spring Uplift in southern Wyoming, on the ancestral lands of the Shoshone-Bannock and Eastern Shoshone.[57] Namely, Landsat data was compared with NS-001 data, and the experiment confirmed that the NS-001 experimental Thematic Mapper was more useful than Landsat's MSS for geological applications.[58] The project NS-001 became part of the Landsat satellites in 1982—with the launch of Landsat 4—the year that the Joint NASA/Geostat Project concluded.

In 1979, Landsat was transferred to NOAA, and with the Land Remote Sensing Commercialization Act of 1984, the Landsat system was commercialized. Comsat, a "quasi-private firm chartered by the federal government," had hoped to commercialize US weather satellites simultaneously in order to sell weather data back to the government and therefore finance Landsat.[59] Despite this desire, the act prohibited "the sale of any part of the meteorological satellite system to private industry, thereby reaffirming that it [weather data] would remain in the public sector and that the data would be collected, archived and disseminated by public agencies freely and free of charge to users on an open and non-discriminatory basis."[60] While the fate of weather satellites and their data would be different than the fate of Landsat, several key points about their overlaps are clear. First, Landsat, a satellite that grew out of NIMBUS and TIROS, was key to the development of the extractive industries in the United States. Second, Landsat's evolution reveals the growing tension between environmental data as a public versus a proprietary good.

Weather data—much like resource data—was a hot commodity within the private sector and was of just as much interest to oil and gas companies as Landsat data was. By the 1980s, NASA had launched seven NIMBUS satellites as well as over twenty-eight satellites under the TIROS program.[61] The data from these satellites were available to the public as well as to private industry. Meteorological data became important for the oil and gas industry in a variety of ways. Petroleum "jobbers"—purchasers of refined fuel from refining companies, who would then sell that fuel at retail price—relied on temperature prediction to anticipate a consumer's fuel usage and deliver oil when fuel tanks are low.[62] Offshore oil and gas operations relied on private meteorologists for wind and wave forecasts, as well as for meteorological forecasts to inform them when they should move their rigs.[63] These data were also valuable for insurance companies as they could utilize private meteorologists to assign risk related to the liability of weather damage.[64]

Just as land-based settler weather data collection facilitated Indigenous land theft in the nineteenth century, the utilization of satellite data for the proliferation of oil infrastructure has directly impacted the experiences and land realities of Indigenous communities in the late twentieth and early twenty-first centuries. Landsat satellites, for example, have been used to execute conceptual site plans, as well as habitat and impact assessments, for several oil pipelines that cross Indigenous land and threaten land sovereignty. These include Keystone XL, Dakota Access Pipeline, and Line 3. For the Dakota Access Pipeline—which was proposed to cross the land of the Standing Rock Sioux people—the company Geoengineers utilized Landsat 8 data via Google Earth to execute conceptual site plans and profiles for the project.[65] The Keystone XL oil pipeline project utilized Landsat data to perform habitat assessments to affirm pipeline viability.[66] Keystone XL's proposed path crossed the homelands of Great Sioux Nation and the Gros Ventre and Assiniboine tribes.[67] Line 3 utilized satellite data for an impact study assessment of the pipeline route that passes through the land of the Anishinaabe.[68]

While the private sector leveraged earth's weather and resource data to gain market power, more and more calibrated and cleaned, gridded, and superimposed weather data products made their way to audiences near and far. Weather telecasting, as Robert Henson has argued, put "raw data in perspective."[69] In the mid-1960s, TIROS satellite images were aired on two Florida stations—Jacksonville's WJXT and Tampa's WTVT—and New York's WNBC. As Henson elaborates, "Satellite pictures became an integral part of local weathercasting starting in the late 1970s with Geostationary Operational Environmental Satellites."[70] GOES hovered in the same position and was able to send satellite pictures back to ground receiving stations every hour, eliminating "data gaps" and allowing for "a full day's set of photos" to be "assembled into dramatic satellite loops."[71] By the 1970s, major news networks received daily satellite loops from NOAA, and by the late 1970s, "most large-market stations had incorporated satellite pictures into their weather segments."[72]

Henson observes that GOES made "satellite pictures . . . an integral part of local weather telecasting in the late 1970s."[73] Calling "the appeal of satellites to viewers . . . obvious," he explains, "People could see weather occurring across the entire United States at once. Moreover, satellite images had an edge over radar displays in that they showed the tops of clouds whether or not they were producing rain or snow, thus providing a stronger visual element on dry days. Satellite loops also helped reveal how day-to-day weather evolves. The explosive growth of thunderstorms, with their bright images and rapid motion, could easily be contrasted with less dramatic cloud formations hovering elsewhere."[74] Satellite images, as used by television networks, both "reveal[ed]" the workings of everyday weather and offered up "dramatic formations" of what was happening above viewers' heads. Most important, they always revealed these workings in a visual way, placing an important emphasis on vision as a form of weather knowledge.

GOES AND HURRICANE SANDY

Three days before Hurricane Sandy's East Coast landfall, NASA released its first aerial, time-lapse video of the storm.[75] Roughly twenty-two thousand miles above the earth, NOAA's GOES 14 satellite—the twenty-first-century successor to the TIROS fleet—scanned the hurricane's movement across the southern United States. The first "movie" (a composite of rapid-scan images), released on October 26, 2012, captures the enormity of the storm, its dynamic cloud structure, and the sharpening of the storm's eye as the continent dims under the setting sun.

Hurricane Sandy was massively mediated, making it the perfect storm for satellite meteorological data.[76] It was a "weird global media event," as McKenzie Wark describes, in that the hurricane was a storm both in the sea and across a myriad of media vectors. Sandy happened, to borrow Wark's language, "within a space and time saturated with media"; the storm was "weird" and "global" in the sense that the event "travers[ed] borders" and "call[ed] a world into being."[77] I use Wark's term *global media event* to classify Hurricane Sandy because it signals the spectral life of mediated weather crises in the United States and beyond: the ways these crises call another world into being that acts not as a rendering but as a reality. These realities are fueled by meteorological data. The flooding of lower Manhattan and the electrical outage below Fourteenth Street, the drownings in Midland Beach, Staten Island—these events, among others such as fires in Queens and massive flooding along the New Jersey coast, both confirmed the reality of predictive severe storm data (yes, the water rose that high) and also baffled and exceeded the imaginaries that inhabitants of these spaces held about extreme weather itself. As New Jersey resident Patricia Armstrong remarked, "I couldn't believe the water had gotten so high."[78]

While NASA's GOES 14 satellite is a fundamentally more advanced tool for sensing earth weather than the satellite that captured the 1960 TIROS photographic scans of earth, the popularity of GOES images during Hurricane Sandy—as this next section will explore—exemplifies the extent to which, almost fifty-five years after TIROS's launch, institutions and publics are still uniquely drawn to visual and remote weather perspectives.[79] More precisely, TIROS inaugurated, and GOES participates in, an age of weather sensing and visualization that depends not only on the availability of wide fields of environmental data but also, more precisely, on the sovereignty of visual data.

The satellites GOES 1 through 3 were spin-stabilized satellites that moved at 100 RPM. They were able to view earth only 10 percent of the time, and their role was to monitor extreme weather events through a visible infrared spin scan radiometer, which provided full-disk imagery of cloud conditions during the day and at night.[80] GOES 4 through 7, which were operational from 1980 to 1996, added sounding technology—vertical profiling of moisture and temperature—a welcome improvement to the imaging technology that profiled cloud conditions. Unfortunately, the sounder and the imager could not operate simultaneously, making it difficult to gather cloud and temperature data for storm prediction. This changed in 1994 with GOES 8 through 13, all of which were equipped with simultaneous and independent imaging and sounding devices. These instruments not only were able to operate at the same time but also were able to operate 100 percent of the time, thereby offering constant data streams. These satellites also were able to receive data transmissions from on-the-ground data collectors, such as balloons, buoys, and other remote, land-based automated data collection stations.

The GOES 14 meteorological satellite was not only capable of constant data streams like its predecessors but also was the most powerful geostationary satellite to date when it was launched in August 2012. The GOES 14 satellite began as an experimental venture in rapid scanning and was used as an emergency satellite to collect rapidly scanned images of Hurricane Sandy. The satellite's rapid scans were collected and communicated to ground-based stations every minute, as opposed to the scans of previous satellites that were received every fifteen to thirty minutes.[81] Rapid scanning is embedded in centuries-old settler knowledges about the environment, weather, and climate and is a particularly salient form of narrative for elaborating meteorological data. As a form of data collection, it shows both what is deemed sacred and what is deemed disposable. Hurricane Sandy was a novel storm event not only because of the damage it inflicted on some of the most populated spaces in the United States and Puerto Rico but also because it released rapid scanning, a new technology for narrating weather events, into scientific and public consciousness.[82]

The fact that the satellite was even available to record the storm was, historically speaking, an accident. Launched on August 16, GOES 14 finished its intended experimental rapid-scan phase on September 24, 2012. It was slotted to be placed back in "storage mode" orbiting the earth in late October. But when researchers at NOAA, the National Environmental Satellite, Data, and Information Service, and the Cooperative Institute for Meteorological Satellite Studies reviewed track forecasts that suggested Sandy would collide with the East Coast, a special request was sent to satellite operations to save GOES 14 from storage and have the satellite resume its rapid-scan activity during the development and landfall of the storm. When NOAA began its experiment with GOES 14, the association had no idea that its experimentation period would document six days of the storm.

Although the process of obtaining a "movie" of Hurricane Sandy might have felt seamless to publics on the receiving end, rapid scans (and the loops or movies that can be made from them) are processed materials that begin, in their most basic form, as uncalibrated, messy constellations of data. The meteorological movie is created from lines of binary code gathered by a network of noncontact, remote sensing, and in situ instruments, including a two-axis mirror scan system. It looks like a "motion picture" because the image data has been calibrated, cleaned, and visualized. In fact, what comes to us in the form of a movie is governmentally categorized as a "product" of the GOES Ingest NOAAPORT Interface, otherwise known as GINI. The movies are produced by a visualization machine that is therefore data driven and data based. According to NOAA, GOES's rapid-scan images allow users to view satellite imagery in real time. Since the scans are separated by only sixty seconds, one could reconstruct a real-time image of earth weather by looping various sixty-second scans together. Rapid scans fill in temporal details that a fifteen- or thirty-minute scan naturally misses; the result is a more dynamic temporal picture of what is happening on earth.

The NASA movie of Hurricane Sandy spread swiftly across the web in the days leading up to the hurricane's East Coast landfall and during the storm's heaviest hours, collecting over one million YouTube views.[83] The movie, composed of "loops," as NASA scientists would call them, appeared on the Weather Channel, CNN, NBC, and Reuters in addition to being studied by NOAA, the Ocean Prediction Center, the Weather Prediction Center, the National Hurricane Center, and the National Weather Service (NWS), eastern and southern regions. The different language used for the "movie" or the "loop" shows how these rapid scans were variously experienced by the audiences viewing them and affected by the context in which they were viewed. In popular culture, the rapid scans were entertainment—"terrifying" and "beautiful" as well as "awesome" in their ability to render the storm.[84] Responses like these from popular sites *Gawker* and *BoingBoing*—responses that depend on a global telecommunications infrastructure—are characteristic of untutored responses to storms of all types. For others, like the researchers at NOAA and the Weather Prediction Center, these were loops full of "data" that, if read properly, could change the science of storm prediction and modeling forever. What the many lives of these media show us is the fullness with which these scans operated as both aesthetically rich media and objects of scientific study before and during the storm event. The scans were beautiful, but they also represented certain scientific truths for researchers.

Satellite media constitute a reality in that satellite technologies structure not only how members of the nation-state and media consumers know "the world" but, more precisely, how they know the world as an "environment"—a planet pulsing with wind, rain, polychlorinated biphenyls, and mercury. The promise of a meteorological satellite is the promise of the extension of vision over settler space and time. Such an extension stabilizes colonization as the beginning of "settler time" and the nation-state as terrain on which the drama of settler "progress" is played.[85] Meteorological satellites certify that to know the weather is to see it as data within the long arc of the nation-state's scientific progress, from standardization to forecasting to computer modeling. However, such a perception of progress fails to incorporate how meteorological satellites are technologies born of domination, surveillance, extraction, warfare, and dispossession. In addition to creating vast swaths of data used for storm monitoring and weather forecasting and prediction, meteorological satellites simultaneously engage these other activities and are part of multiple temporalities in which "progress" can be read as dispossession, racial capitalism, white supremacy, and a thousand climate change apocalypses across Turtle Island beginning with the onset of settler colonialism and expanding into the present day. These "infrastructures of White settler perception," to extend Paul Guernsey's conceptualization, are propelled and proliferated by meteorological satellite media and fail to account for, in the words of Kyle Powys Whyte, the fact that anthropogenic "climate change is an intensification of environmental change imposed on Indigenous peoples by colonialism."[86] The environmental crises displayed back at media consumers through the satellite's eye are new emergencies only to settlers.

As this book has explored, settler utilizations of weather data date back to the mid-nineteenth century and were closely aligned with land theft, genocide, settlement practices, and the acceleration of settler colonialism. Settler scientists wanted to quantify the weather in such a way that would lead to standardized weather data collection practice that could be applied across the US settler states. They were successful in achieving this goal; they translated the weather—the clouds, precipitation, and air density—into numerical weather tables; they compiled, calculated, and shared these tables among members of scientific societies and with farmers, teachers, and librarians with an interest in settler data. Across the twentieth century, quantification grew to include aerial exploration. Precursor technologies that led to meteorological satellites included mountainside barometer readings in the mid-seventeenth century, thermometers placed on kites in the eighteenth and nineteenth centuries, cameras placed on kites in the nineteenth century, weather balloon ascents above fifty-three thousand feet in the late nineteenth century, and finally rocket sensing and airplane photography throughout the twentieth century.

Meteorological satellites are guiding technologies that allow the nation-state's meteorologists and media consumers to see weather as a global phenomenon through the "technologized vision" of the satellite itself. To return to Parks: "Satellites are now at the core of our global telecommunications infrastructure, and . . . have become a principal means by which we see and know the world and the cosmos beyond." Astronomical and earth-facing satellites, Parks argues, are "extending the domain through which distant vision and knowledge are possible." Although satellites occupy the peripheral space surrounding our earth, they significantly influence the ways members of the nation-state can come to know their environmental worlds. The act of witnessing "demilitarize[s] military perspectives"; [87] in other words, it understates the eye of the nation-state through the language of a global scientific view.

Satellite weather data circulates—uncalibrated, calibrated, and cleaned—in a knotty web from GOES *up there* to ground stations *down here* in Wallops Island, Virginia, and Camp Springs, Maryland. The NOAA uses the term *memory* to refer to the computational space in which the soon-to-be processed data lives. When the proper data is "recalled," GINI can form a visual product (in the form of Cartesian graphics like cloud cover, temperature, wind velocity, and precipitation "maps") that can be transmitted to users at weather outlets near and far. But the images are also built *in memory* in the sense that they are recollections of a weather past witnessed only by the satellite's sensors. The memory—its detail and temporal logic—is a reconstruction of the satellite's optical port and telescope mirrors. The transmitted data does not, in the end, constitute a representation of a global scientific view but rather a view unique to white settler perception and the environmental relations encoded within this perception.

It was Thursday, October 25, 2012—just a few weeks before the end of apple season and three days before Hurricane Sandy's East Coast landfall. Houses and ocean-side apartments were adorned with jack-o'-lanterns and vinyl ghosts, in anticipation of Halloween. All along Hunter Avenue, children were making last-minute costume decisions (maybe Katniss Everdeen or something from the new *Ice Age* movie), while adults made quick trips to the nearby ShopRite, hoping to find some candy on the ransacked shelves. It would not be until the following morning, Friday, October 26, that then mayor Michael Bloomberg would give his first public news briefing on the approaching storm, issuing a state of emergency. And it would be roughly three more days before the waters of Lower Bay would spill onto Hunter Avenue, drowning the Halloween decorations, bits of houses—and at least seven neighborhood residents.

The next day, residents of Midland Beach, New York, would watch the news reports play over and over again the silent roll of the storm's eye up the southern seas with a projected landfall in the east. Many would evacuate Midland Beach (an area of Staten Island located in flood zone A, the most likely area to flood), but others would stay, retreating in the heaviest hours of the storm to the very tops of their bungalows, to roofs if they could manage to stand on them despite the wind, or into canoes and boats to battle the water that had turned the streets into waterways. The story of Midland Beach is one of a locally experienced and situated weather crisis, where residents were both the consumers of satellite images—and all the warnings that came along with them—and the victims submerged under the imagery.

Analyzing the traces of Hurricane Sandy and NOAA material in the context of remote sensing history means working with what Wendy Hui Kyong Chun has called the "undead of information," which consists of "reworked" and "revisited" network matter.[88] Although weather is physically experienced as a localized event, the data narratives produced by the visualization of weather systems such as Hurricane Sandy have the capacity to cut across diverse and seemingly separate geographic and temporal spaces, suspending and obscuring boundaries. Weather visualizations like *Wind Map*—as well as media proliferated by NASA and NOAA—have come to shape the level of engagement, the variety and locale of feeling, that consumers of media have with weather, especially extreme weather. During a political and ecological moment in which climate change is increasingly positioned by settler and imperial nations as a universal and global emergency, I seek here to illuminate how the visualization of environments through meteorological data (or the aesthetic space of meteorological data visualizations) is enmeshed in settler worldviews that serve particular ends for the nation-state. The popularity, continued transmission, and contextualization of GOES satellite images across a variety of web platforms manifest a desire to aestheticize these visualizations in a way that obscures the following: the lived scale of weather crisis and the ways ongoing settler colonialism and dispossession are central to lived crisis. Meteorological satellite data is mobilized toward both of these ends.

Weather visualizations—visualizations that circulate as part of a global media event—ostensibly reveal quantitative, data-driven messages about storm realities. But these visualizations separate data from the lived realities of environmental harm and culpability. As a view of and from the nation-state, satellite meteorology is mobilized during crisis events in order to separate "knowledge from responsibility."[89] Doing so ensures that the practice of data gathering (of settler science) can continue without having to address harm and suffering. Quantifying the crisis of the weather is an essential function of satellite technology in times of environmental emergency. Dressed in the programming language of Java, the visualized storm realities become aesthetic products: a picture, a loop, or a movie. They are meant to be consumed and sometimes to entertain. Despite the real-time suffering and the complicated process of data gathering and transmission, these movies come to express pure and artful storm realities. Put another way, the messiness of environmental data (and the nation-state relation to it) seems to melt away when these images circulate as aesthetic—beautiful and terrifying—visions. The beautiful storm eye is beautiful not because it is a "product" of GINI but rather because it appears as seamless as a movie, representing a noninterpretational weather reality whose vision or positionality is not interrogated or held suspect. For that very vision (which was, in the beginning, lines of code and uncalibrated data) escapes status as a representation.

In the early months of 2012, coleaders of Google's "Big Picture" data visualization group, Fernanda Viégas and Martin Wattenberg, released *Wind Map*—an interactive weather visualization project implemented in HTML and Java that visualizes surface wind data from the National Digital Forecast Database (NDFD) (figure 5.1).[90] Unlike more traditional maps, such as NOAA's graphical forecast wind map, Viégas and Wattenberg's *Wind Map* represents wind as a perpetual, fluid body that transcends county and state—though not national—borders in addition to geographic features like mountains, highways, and buildings. The morning of Hurricane Sandy's East Coast landfall, *Wind Map* made its rounds through various outlets, including Twitter, *BoingBoing*, *Wired*, and *Slate*. The home site also had a spike in hits during the days leading up to Sandy and during the hours of Sandy's landfall. Taglines advertised the "Real Time Wind Map of Sandy," described how it "conveys the storm's effects on the East Coast's wind patterns clearly and beautifully," and noted that "the surface wind data in this beautiful wind map comes from the National Digital Forecast Database."[91] To consecrate Sandy's swirling winds, the site archived two GIF images from Sandy's landfall and journey up the East Coast. Graphically enhanced still images extracted from these GIFs are also available for purchase online—$450 for a 32-by-42-inch print. The way *Wind Map* was used during and after Hurricane Sandy suggests that the map's value lies equally in its artistic construction and its indexical nature—the "beautiful" and "clear" map conveys "effects" in "real time" through the implementation of national data. So, the map is pretty to look at, but it also operates as a form of evidence, as an indexical marker of the real. Although data visualizations such as *Wind Map* are not photographs, they operate, in part, like photographs during moments of weather crisis as they become forms of evidence of the storm's movements. Once again, however, these evidentiary representations obscure the lived scale of weather crisis as well as the ways ongoing settler colonialism is central to this crisis. Hurricane Sandy presented challenges to the region's settler infrastructure, including energy, transportation, water and waste, and communications. Climate change and severe weather (including Hurricane Sandy) vulnerabilities to these infrastructures, and the ways of life they support, were known and elaborated prior to Hurricane Sandy's landfall.[92] Yet, knowledge of such vulnerabilities did not drastically reshape established ways of living—settlement patterns and infrastructure development—on the island of New York.

Wind Map gathers data from the NDFD in order to make its real-time visualization. The NDFD itself relies on Weather Service field offices to clean and package meteorological data into matrix entry, graphical entry, and model interpretation form so that it can be entered into the database and accessed by a variety of users in the private and public sectors. Coming full circle, the field office acquires data from a variety of sources, too, including surface observations collected from automated surface-

FIGURE 5.1. *Wind Map*, created by Fernanda Viégas and Martin Wattenberg. Image from October 29, 2013.

observing systems, marine observations collected through weather buoys and weather ships, upper-air observations collected through weather balloons equipped with radiosondes, and meteorological satellites. This is all to say that the data being pumped into the NDFD comes from many different sources and scales. What we see as a real-time visualization of wind over the continental United States through *Wind Map* is not only an amalgamation of many forms of data collected at a variety of scales but also, most important, a real-time forecast rather than a real-time representation of the movement of wind, subject to the uncertainty and error of human prediction. Popularly, *Wind Map* garners an indexical status similar to that of the photograph. It operates as a "certificate of presence," as Roland Barthes once wrote, as "authentication itself."[93] Reception of the map suggests that the visualization operates as an imprint of reality precisely because it works with real-time data, just as photographs attempt to reproduce live or real-time subjects. But the predictive quality of the map throws it into sharp relief as a representation of weather rather than a pure presentation. *Wind Map* is a visualization in the sense that it works, in the same way that NOAA's loops or movies of Sandy did, with inherently nonvisual data sets. And the product is a visualization because it works with "predictive" data, or data that dreams of potential weather realities, rather than captures them.

"A way of certifying experience," Susan Sontag suggests, "taking photographs is also a way of refusing it—by limiting experience to a search for the photogenic."[94] Wondering what the photograph excludes from its frame, Sontag recasts Barthes's undeniable "certificate of presence" as only a fragment of the real. While photographs, for her, are devoid of interpretation, which is why they need captions, Judith Butler unpacks the way photographs offer an interpretation of the world by "framing reality," through which "the photograph has already determined what will count with the frame—and this act of delimitation is surely interpretative, as are, potentially, the various effects of angle, focus, light, etc."[95] In its attempt to visualize the real, photography determines how the real should appear not simply by limiting itself to photogenic content but also by structuring its vision along particular angle and light values. We need to bring that same attention to environmental visualizations such as NOAA's satellite movies and *Wind Map*. Weather satellite imaging and weather data visualizations (which borrow aesthetic devices from cartographic weather products and photographic imaging) build on the history of visual forms such as photography as well as the complicated history of graphical presentation. Since weather is a dynamic system that more or less refuses absolute visual and static presence, visualizing and rendering weather will always be a complicated task imbued with a complex set of back-end techniques, including the invention of a spatial imagination and, more concretely, georectification, or the process of aligning a map with a spatial grid.

The problem that consumers of weather culture and media face is that the back-end processes never pronounce themselves. In other words, the visualization arrives as a purity, as an uninterpreted presentation of environmental realities. As Johanna Drucker writes in *Graphesis*, "Data pass themselves off as mere descriptions of a priori conditions. Rendering observation (the act of creating a statistical, empirical, or subjective account or image) as if it were the same as the phenomenal world and its interpretation, undoing the concept of interpretation on which humanistic knowledge production is based."[96] Data is constantly trying to skip the acknowledgment that it is an interpretation, as a curatorial vision made possible by social, cultural, and technological conditions. This is not to devalue data itself—data is valuable, important, lifesaving—but it is to reorient our gaze at environmental data visualizations so that we can see, as Drucker phrases it, the "interpretive construction."[97]

Movies produced by NOAA and NASA, as well as projects like *Wind Map*, provide interpretations of environmental realities that are built on vast sets of environmental data. During storm crises, these data visions travel without status as interpretation and often within the discourse of the aesthetic and sublime. They also comply with statistician Edward Tufte's rules for graphical excellence. Strive toward "varying shades of gray," Tufte tells us, rather than large quantities of color that "generate graphical puzzles."[98] Aspire to "simplicity of design."[99] *Wind Map* does this by maintaining a singular lineation scheme. Rather than visualizing the mapped lines of roadways and state borders or the spread of carbon monoxide and particulate matter, it displays only wind lines. The NOAA and NASA images also work toward simplicity, presenting themselves as movies or photographs rather than the multilayered, uncalibrated data composites they begin as.

In an age of weather satellites, weather "data" becomes synonymous with photograph, movie, or map. Such curated imagery is popular and transmittable not only because it has indexical value as a product of "real time," apparently photographic and photographable data, but also because it capitalizes on popular visual taste: simplicity and elegance in design, or simplicity and elegance in the expression of weather realities. But simplicity is a complicated aesthetic value when it comes to US environmental realities, and especially legacies of toxicity and histories of environmental justice. If *Wind Map* is a photogenic visualization—which explains its popularity—and if it aspires toward graphical excellence, as Tufte would say, it is because it leaves unvisualized ugly or unpalatable data sets. What would happen if a visualization like *Wind Map* had to account for ugly environmental data while still striving toward photogenic or graphical excellence? How many people would buy a still image of *Wind Map* if it depicted sewage overflows or the distribution of volatile particulate matter that travels along with the wind?

Distinctly new productive technologies for mediating the environment in general and weather specifically, including large data collection and imaging tools such as the GOES satellite, open-access databases such as the NDFD, and project implementations of these datasets through HTML, JavaScript, and other programming languages, as *Wind Map* shows us, have repurposed how consumers understand data and the weather. The repurposing of NASA's GOES satellite imagery—the design and digital life of *Wind Map*—highlights a popular way of seeing weather in the age of climate crisis: from a distance and without much attention to ugly data realities that implicate the nation-state and its settlers in the making of disaster. Rather than drawing attention to the dimensions of climate change and weather crisis, these projects obscure the sensorial and affective tendons that bind consumers to their geological and atmospheric realities. They cleave the land from the flesh.[100] This lack of connectivity makes it easier to position weather disasters as purely "natural" events rather than overwhelmingly anthropogenic and toxic ones.

Satellite images are highly calibrated and cleaned data products built through access to more data than scientists have ever had. Yet their views are based in a visuality that removes actors from the extreme weather equation. Meteorological satellites have a long history of collecting data used for extractive ends, deterritorializing human and biotic communities. The history of weather satellites' relation to military reconnaissance, natural resource monitoring, and extraction is a story of settler domination mobilized by weather data. Caring for the construction of a pipeline through data collection and management is not the same as caring for rights of collective continuance as it applies to human and biotic communities, waterways, air, and soil. The logic of the satellite's eye is steeped in settlement, extraction, domination, and strange settler pleasure through the aesthetics of environmental horror. Perhaps rather than caring for data—hungering for more, and more—we need to ask and answer: Whose survival is implicated in the type of data meteorological satellites collect, produce, and share, and whose is not? As a way of knowing environments, data builds relations, but the relations data can build are always up for modulation.

Hurricane Sandy was more than a crisis of infrastructure; it was a crisis of the environmental imagination, to borrow a term from ecocritic Lawrence Buell.[101] More precisely, it was a crisis of the settler environmental imagination, one that has been founded on domination and capture as primary ways to relate to weather and environment. "I just couldn't believe the water had gotten that high," said Patricia Armstrong, a resident of Seaside Park, New Jersey.[102] And Armstrong was not the only one who was shocked. Interviews and oral histories conducted after Sandy reveal a common experience of disbelief across the Eastern Seaboard.[103] Buell writes that "the environmental crisis involves a crisis of the imagination, the amelioration of which depends on finding better ways of imaging nature and humanity's relation to it."[104] Although I disagree with Buell's language of "humanity" here—especially given this book's focus on settler relations and responsibility—his observation helps us contemplate the circuitous ties between the material environmental crisis and the environmental imaginaries that emerge somewhere between the experience and its depictions in popular media. Although Buell was writing before the popular rise of weather data visualizations, he is right to say that we need better ways of "imaging." Imaging, after all, is at its core a form of imagining.

The visual response to Hurricane Sandy manifests not a failure at "imaging" better relations between the US settler state and nature but a failure to imagine any affective relation at all. As this book has intended to show, settler science has curated a relationship to meteorological data founded on dispossession, domination, and control. What have we, members of the nation-state, inherited from these historic relationships? In quantifying the weather, settler science measured its way out of reciprocal relation. Feeling the fewest impacts of climate change and extreme weather, the US settler state and its residents have continued to accelerate it—shaping intense storms like Hurricane Sandy—and suffer from collective forgetting regarding contemporary environmental violence, historic dispossession, and culpability for an imperiled biosphere.[105] The visual response to Hurricane Sandy magnified a tear in this already broken network of relations.

THE DATA OF CLIMATE CRISIS

Encompassing the rise and proliferation of US satellite meteorology, the final chapter of this book has traced the historic interdependence of weather data collection via satellite meteorology and US settler colonial logics. The first weather satellite, TIROS, was a product of a US reconnaissance concept originally developed by the RAND Corporation in consultation with the US government. After it was built and launched, the actual TIROS satellite paved the way for NIMBUS and, later, Landsat—a satellite used by the extraction industry for oil and gas production. Across the middle to late twentieth century, satellite vision—and the meteorological data that fueled it—became a central part of how the nation-state saw itself, and how it was seen by its inhabitants.

Imbued with the modalities of information capture that were central to the functioning of the nation-state—such as capacities to sense weather patterns, mineral and fossil deposits, and enemy combatants—satellite vision also came to shape popular understandings of the weather for those who viewed and consumed satellite media. In the early twenty-first century, satellite vision and its data have become central to how climate crises are narrated by a nation-state responsible for much of the fossil fuel extraction and burning, waste production and use, and chemical pollutants the biosphere struggles to survive under. The GOES 14 project's rapid scanning—an advanced technology that grew out of less advanced scanning technology on earlier versions of GOES—measures environmental crises in seconds, leaving no intended field of capture unscanned. The dream of continuous data had been achieved, but what rotten worlds has it left in its wake?

The overabundance of data has made little difference in communicating to the nation-state and its inhabitants the depth and severity of the problem at hand, suggesting that more data will not save "humanity" —it is not humanity that needs saving. Much like Heather Davis and Zoe Todd, Andreas Malm and Alf Hornborg point out that the Anthropocene is not a product of humanity in general but is rather the product of a particular segment of humanity that built and participated in European colonization. The "fossil economy"—one of the drivers of the Anthropocene—"was not created nor is it upheld by humankind in general," and therefore an understanding of cause must be linked to "intraspecies inequalities."[106] Lawrence Gross contributes that "Native Americans have seen the end of their respective worlds," and Kyle Powys Whyte adds that "climate change is an intensification of environmental change imposed on Indigenous peoples by colonialism."[107] Indigenous peoples are not the creators of the Anthropocene or climate change but have, rather, suffered and survived the apocalypses introduced by settler culture. The climate crisis—including the slice of crisis narrated by GOES data during Hurricane Sandy—is not a crisis of "human" making. It is a crisis of the US settler state and other imperial nations that have measured, extracted, and plundered the biosphere in the name of capital accumulation, sustaining power, wealth, and extractive ways of life.

To know the weather through satellite meteorological data means to face the very settler knowledges that have abstracted settlers from environments in the first place. The nation-state's claim to know the land—and to seek to dominate it and control it—through data collection is its progress and peril. Through data, we have made an archive and an atrocity of the weather. We claim to know the biosphere through numerical capture yet fail to know that same biosphere any other way, especially in vulnerability and need. We refuse to know the biosphere as a network of living beings, each with their own reasons for continuance that exist outside and often against our own desires. We refuse to see our own implication in and responsibility for the cruel deaths served to this living network and the myriad communities that depend on such for care, sustenance, and survival, including ourselves.

The use of meteorological satellite data functions precisely as it is supposed to during weather crises such as Hurricane Sandy. It extracts the US settler state and its inhabitants as actors in the national and global horrors of an avoidable climate change. It absolves responsibility for crimes in the name of objectivity. In this way, satellite vision accelerates the accumulation of settler property and control, which began with nineteenth-century land theft. Yet, at the same time, the weather crises that satellite vision attempts to document—crises accelerated by the uneven and unpredictable effects of global warming—are also destabilizing events within setter colonial timelines in that they challenge settlers' capacity to quantify and claim the biosphere through prediction and data production. Data does not simply attempt to explain; as this book has aimed to show, it also attempts to claim. And when it fails to do so, it can signal an otherwise world, which extends outside of settler systems of knowing.

But how exactly do we get to the otherwise? What comes next, and where can we go from here? By "we," I do not mean all of humanity. Instead, I am speaking specifically from the position that is most familiar to me and to the research of this book: by "we," I mean settlers. And the otherwise I am concerned with is a future in which the violence of settler colonialism no longer structures everyday life. Settler responsibility and decolonization (the material process of repatriation of land and Indigenous sovereignty) are central to the creation of this future.

For settlers to be accountable to the past and present—and to let go of the reckless settler futures that promise endless plundering of the earth—they first must know settler history, become accountable to that history, and become accountable to the communities that have been harmed across that history. In this book, I have argued that settlers have utilized notions of data to disentangle themselves from networks of relationality, responsibility, and care across the nineteenth, twentieth, and early twenty-first centuries. Settler science was never about accountability; it was about appetites for environmental knowledge, and the lands this knowledge grew within. What would becoming accountable to settler environmental history require? It would, in the words of Rebekah Garrison, require working against "imperial amnesia," which constructs history "not through Indigenous cultural memory but through the stories that settlers tell about the past," which "are foundational to this US historical erasure and the proliferation of settler ignorance about US History."[108] Garrison emphasizes the way that imperial amnesia obfuscates settler responsibility by "den[ying] the existence of counter narratives" that would require accountability for ecological destruction and genocide. In doing history, Garrison calls for a version of settler responsibility that is place based and that foregrounds critical self-reflection both of imperial amnesias within settler communities and of solidarities between Indigenous and non-native communities.

I agree with Garrison that settler responsibility is a necessary next step. However, the question of solidarities is a complicated one. Settlers often desire solidarities with Indigenous communities, yet these solidarities are not necessarily theirs to initiate and choose. Centuries of trauma and harm cannot simply be undone through the wish and attempted practice of solidarity; nor should historically harmed communities be the recipients of solidarities they do not desire. That, after all, is not really a solidarity but rather a settler dominance and privilege rearing its head.

The framework of restorative justice, as well as transformative justice, offers a way to attend to the question of solidarity anew. The move toward solidarity is premised on the recognition of the historical and contemporary conflict caused by settler colonialism. Restorative justice recognizes conflict as a generative moment, "an opportunity to learn something about each other and the systems we've created together." Restorative justice begins by recognizing Garrison's "imperial amnesia" ("investigating the systems we've created together"). Importantly, however, "restorative justice also doesn't mean that all relationships can be restored after harm has happened . . . complete recovery of a relationship isn't possible or necessarily desirable."[109] Settler moves toward solidarity and building relationships with Indigenous communities need to account for whether historically harmed communities are desirous of relationships with settlers; such moves need to be grounded in an understanding of restorative justice that does not center settler needs or desires for comfort or assuagement of historical guilt, but rather give energy to the needs, boundaries, and desires for recovery of a relationship from the perspective of those who have been harmed. Reconciliation is not always possible or wanted by those who have been harmed, especially when such reconciliation, as Tuck and Yang write, "is about rescuing settler normalcy, about rescuing a settler future."[110]

Settler responsibility and engaging restorative, transformative justice are necessary steps toward decolonization. Importantly, settler respon-

sibility requires not only confronting imperial amnesia but also the dissolution of settler ways that are founded on domination, extraction, and
genocide. This dissolution—"transforming the conditions that make
injustice possible"—is central to transformative justice.[111] Decolonization might be thought of as a form of transformative justice, which
involves, as Avril Bell and coauthors have articulated, "undoing the structures of settler colonialism" and "replacing relations of domination and
appropriation with new relations of engagement and mutuality"—this
being a "transformation of the settler being."[112] Such a transformation of
being is needed and necessary, but it must also be accountable to historical harm and the intergenerational traumas that have been perpetuated.
Settlers cannot enter new relations of engagement and mutuality without accountability, which in restorative justice "is the ability to recognize,
end and take responsibility for violence."[113] While decolonization "is not
accountable to settlers, or settler futurity," it is accountable to "Native
futures, the lives to be lived once the settler nation is gone," and therefore
may also be accountable to those lives that have undergone transformation by actively undoing the structures of settler colonialism and taking
responsibility for generations of violence.[114] There is space for so many
of us in this future, if only we are willing and able to end structures of
harm, and the identities they have informed, since this nation's founding.

Standing at the coast in 2019, I watch the moon melt onto the ocean. It's one of those August moons—golden as it rises above the water, almost like an alternate sun. Under simile, moon becomes sun and the world shifts, collapses, until another blooms anew.

Standing at the same coast and moon in 2022, grief festers and spills out. Personal grievances crowd our personal oceans like a full moon's light, and our collective grievances become the air we breathe—grief for what we have inherited, for what we have seen and lived with but not changed, for what the simile promises but cannot name. Grief is a place without solid ground. Grief is a climate unbound.

In the process of writing this book, the climate crisis has worsened, the COVID-19 pandemic has ravaged global communities and especially high-risk disabled folks, and so many other daily violences—environmental but also otherwise—have brought the worlds we inhabit to pieces. True, our worlds were already in many pieces, but as I look around, I find myself struggling to imagine a new arrangement. I am trying to find a simile, to turn moon to sun, but nothing comes. The weather is getting worse. There is so much to grieve for—grief for lost species, lost lives, foreclosed futures, and unlived pasts. How do we choose? Must we?

In adrienne maree brown's novel *Grievers*, grief is a way of life for the protagonist Dune, who attends to the dead and dying across Detroit as an illness, Syndrome H-*8, shuts down the city, alters daily life, and takes the life of Dune's mother, Kama. Syndrome H-*8 seems to affect only the Black community in Detroit, and as the novel moves forward, Dune learns that the onset of the disease is tied to grief itself—those who suffer and die from the mysterious illness have experienced grief elsewhere in their lives. Grief kills.

While many flee the city, Dune stays to attend to the dead and the dying. In doing so, Dune's character shows what it looks like to live amid and survive unfathomable grief. As the novel shows, grief work is never simply about what is lost—although this is central to grieving—but is also about what is yet to come. Though the novel does not explore what comes next, it emphasizes that the underside of grief can be a new leaf.

ABOVE ME IN THE COASTAL SKY, seventeen geostationary operational satellites hover in orbit around the earth, moons in their own right. Two of them are actively sending data to earth: GOES 17 (operational as GOES West) and GOES 16 (operational as GOES East). The rest of the satellites have been decommissioned.[1]

Meteorological data systems become obsolete. They spoil. The obsolescence of a planetary data system begins with the decommissioning of a satellite, which can meet one of two fates. It is either deposited in the outer orbits of earth, drifting in endless rotation, or it is invited back into earth's atmosphere, eventually burning up upon reentry, a burn that leads to a graveyard in the middle of the Pacific Ocean, what space agencies call the South Pacific Ocean Uninhabited Area. As of 2016, over 250 space items—from defunct stations to satellites—lay in this zone.[2] Of the now seventeen GOES satellites, none have found their fate in the graveyard Pacific. Instead, they orbit endlessly around the earth.

At the edge of the ocean, I think of the GOES satellites—all seventeen of them—in their eternal constellations, each an archive of some former data life. While decommissioned satellites can be understood as literal interstellar and atmospheric waste, they are also markers of wasted, spent, or defunct data systems and the frameworks that brought them into existence. In my reading of the sky and the planetary data systems that roam up there, I see decommissioned satellites as relics of data collection, processing, and storing worlds that have become obsolete, marked as wasteful. They are lost and a loss.

When GOES 1 through 3 entered their graveyard orbits, it was because they were not able to perform vertical sounding, a technique that was made available by GOES 4 through 7. And although GOES 4 through 7 could perform vertical sounding, these satellites were eventually retired into their graveyard orbits because they were not able to perform imaging and sounding simultaneously. The satellites 8 through 12 would take up this task, operating imaging and sounding at the same time, though they would also move into graveyard orbit as sensing and imaging technologies increased. The current geostationary meteorological satellites that provide critical weather and climate information for the United

States are currently GOES 16 and 17. This means that rocket bodies and payloads for fifteen additional GOES satellites are currently rotating around the earth.

Data transmission technology also advanced with each new satellite. From GOES 8 on, geostationary meteorological satellites not only were able to scan earth in a variety of capacities but also were able to absorb land-based station data and coordinate that data with geostationary data. The fourth GOES, for example, "relay[ed] data from more than 10,000 surface locations into a central processing center." Even more than that, as data collection processes improved, GOES satellites moved from WEFAX data format service to low-rate information transmission digital service and GOES variable format. As data formats changed, so did the archives of data that were available to NOAA, NASA, allied scientists, and general publics. From TIROS to GOES, meteorological satellites were good at transmitting the data they sensed—and they got increasingly better at this over time, as GOES 8 and on show—but the satellites were not meant to be storage mechanisms for this data. Where was all of this data stored, and what entity would be responsible for storing it? Meteorological data has always needed caretakers.

When the first TIROS was launched in April 1960—a precursor to GOES—the National Environmental Satellite Service was primarily responsible for receiving meteorological satellite data at two ground stations, one in Virginia and the other in Alaska.[3] In the early days of satellite meteorology, vidicon data was stored on 35 mm microfilm at the US National Weather Records Center.[4] But as the nature of data shifted from televisual data to data sets that were not necessarily or only image based, issues around data storage emerged.

In 1982, the National Research Council (US) and Space Science Board reported that NOAA had been experimenting with mass data storage and retrieval using an "Ampex terabit memory system to store NOAA satellite data."[5] A set of experiments at the University of Wisconsin had allowed a modified Sony video recorder to store mass amounts of "digital GOES data." Several other universities that were utilizing data from GOES through land-receiving stations purchased this device from the University of Wisconsin.

At this time, anyone who had set up a satellite retrieving station could tune in to the S-band and pick up real-time data from the GOES fleet. But long-term data was harder to find. Despite the fact that the National Environmental Satellite Center had been thinking about archiving mete-

orological satellite data since 1970 and was working through the climatological applications of such data, it lacked a long-term archive. The GOES
satellites were doing well tracking the weather, but this data—until it
could be properly stored—could not speak to climate variation or climate change.

Data's inheritance requires material space.

With each graveyard satellite, data collection and transmission capacities increased, but data archiving and storage capacities did not.

In 2002, the George W. Bush administration established the Climate
Change Science Program, which was an effort toward "expand[ing] observations, monitoring, and data system capabilities" in order to "increase
confidence" and "understanding" around "why climate is changing."[6] In
its 2005 "Scientific Data Stewardship (SDS) Implementation Plan," NOAA
began to imagine ways of implementing some of these changes, writing
that it hoped to "reduce uncertainties in climate science" by "creat[ing]
high-quality, operational long-term data sets of climate conditions"
based on satellite and ground observations.[7]

While emerging archives of meteorological data were beginning
to extend beyond the thirty-year climate normal period, NOAA was
attempting to store and make accessible "tens of terabytes of data a day
from satellites, radars, ships, weather models, and other sources." There
was simply too much data to store by a singular entity, or in a singularly
formatted system. Difficulties in user downloads and accessibility of data
led NOAA to create the NOAA Big Data Project in 2015 "to explore the
potential benefits of storing copies of key observations . . . in the Cloud
to allow computing directly on the data." One can imagine the ways this
would be useful for researchers and experimenters alike, especially considering that the equipment for picking up S-band data from the current
GOES fleet has grown more expensive as data outputs for these fleets
have increased.

Interglacial data systems like the GOES satellite fleet speak back to
the once-terrestrial fleet of weather observers who used their senses to
measure the face of the sky, who kept "the top of a chimney between the
eye and any remarkable point of cloud." The eye of an observer, body still,
gazing toward a chimney and a stratocumulus. The eye of the storm looking back to the eye of the spectrometer.

Data's inheritance costs.

DWELLERS OF THIS WORLD have inherited a sky of interstellar refuse, among other things. At the ocean's edge, I grieve the downed satellites at the bottom of the ocean and those that drift endlessly around the earth. This is not grief for an empire but grief for the worlds these machines made and unmade through their data. This is grief for all they have seen: those lost in the pandemic, the ongoing traumas of climate change. What do we inherit from these defunct data streams, from the many layers of material infrastructures, data formats, forms of colonization, dispossession, and power that led to their creation and their demise?

Grievances are losses "like half-torn maps," writes Meg Cass. "Grief gathers in our lungs, the words for everything trapped inside us, insects flailing in the fly-tape of our throats."[8] Grief, as Cass elaborates, can be grave, but I am also compelled to think of grief as the refuse that nourishes and allows for new life. The question is what sort of new life will grow in grief's wake. Data can be used to map the end of one world that beckons and breaks toward a new one; the underside of grief can be a new leaf, just as an August moon melts into sun above water and becomes a new world we can barely see, though we can feel its heat.

SATELLITE DATA SYSTEMS, like GOES and the Moderate Resolution Imaging Spectroradiometer (MODIS), are products of interglacial times, and yet they document, too, with clarity an interglacial way of life—the breakage of ice and the geographies of a melting world. Interglacial data is essential data—data that can be used to map what comes next, to borrow from poet Joy Harjo.[9]

Currently, NOAA partners with Amazon Web Services, IBM, Google Cloud, and Open Commons Consortium as hosts of historical climate data, including data like that from historical satellite archives such as TIROS and NEXRAD, which was once available only on videotape. These data sets are open to anyone, and that means they are open not only to researchers but also to economists, data brokers, and insurance companies.

This history of meteorological data has taught me that more data access does not necessarily mean a just transition or a livable future. Despite data's overabundance, 64 percent of US adults surveyed report that they "never" discuss climate change.[10]

The framing of data as interglacial takes us into the long history of settler knowledge-making. This system of knowledge-making—atomistic at its core—is one of the foundational logics that undergirds landscapes of environmental violence historically and today. The enumerative strategy did exactly what it was designed to do. It turned the biosphere into an object, violated the collective continuance of communities whose lives have been historically and are contemporarily bound to that biosphere, and provided a new way of knowing environments founded on quantification of weather and claims of settler sovereignty over biospheric life.

THE HISTORY OF WEATHER DATA recording within a settler colonial context suggests how settler colonialism, white supremacy, and hetero-patriarchal power informed, upheld, and bolstered meteorological science and data collection in the United States from the early nineteenth century to the present day. As Maile Arvin, Eve Tuck, and Angie Morrill have argued in "Decolonizing Feminism," "settler colonialism has been and continues to be a gendered process."[11] Throughout its chapters, this book has aimed to provide evidence of this as well. I have spent much time reflecting on the negative power of silence within predominantly white, settler histories of science. In failing to name whiteness and settler culture within this history as fundamental frameworks of power, practitioners (from scientists to environmental activists and storytellers) have chosen to remain unaware of the legacies of violence that undergird their pasts and inform their presents. This collectivized forgetting is a powerful logic of the settler state and the racial and gendered hierarchies it creates. The forgetting is bereftness and is as violent as the legacies it refuses to remember.

Historic weather data measurement, recording, and archiving informed and built settler worlds, worlds founded on understanding and relating to environments through systems of numerical capture, categorical assignment, and proprietary relationships to the biosphere. These relationships not only emphasized environments as property but also positioned white male bodies as idealized weather data producers and knowledge makers. This relation both required and upheld white settler colonial claims to land across more than two centuries. It was never just about measuring the face of the sky.

Settlers crafted a national concept of data, and this concept organized settler worlds around the quantification of environments, leaving little room for relations of reciprocity and responsibility with the biospheric worlds they were indeed a part of. Once the concept of data was functional in the later part of the nineteenth century, settlers used it to further settler infrastructure and enact settler futures. The twentieth century accelerated the nation-state's use of data across newly formalized government sectors—such as the Weather Bureau, the Department of Agriculture, the Department of Commerce, the War Department, and the Department of Defense—while relying on observers (both employed and cooperative) to provide meteorological data for the necessary functioning of the nation-state. In the late twentieth century, private and government-funded weather services fueled public appetites for meteorological data graphics. These graphics, however, further distanced consumers of media from the crises—the extreme weather of climate change—they observed through data products.

This book documents the two-century wake of settler data meteorology and the worlds and worldviews that have been inherited from efforts to measure the face of the sky.

IN NOVEMBER 2019, the Intergovernmental Panel on Climate Change released another report, this time on the "emissions gap." The report tracks progress toward globally agreed-upon goals. The report provides data. It focuses on bridging the gap between current emissions and projected targets. "The summary findings are bleak," the introduction states. Despite their access to data, "countries collectively failed to stop the growth in global GHG [greenhouse gas] emissions, meaning deeper and faster cuts are now required." "Decarbonizing the global economy will require fundamental structural changes," the report reads.[12] If effectively implemented . . . this would lead to . . .

This would lead to . . .

This would lead to . . .

INTRODUCTION

1. Long Soldier, "Women and Standing Rock."

2. What is now called Philadelphia sits on Lenapehoking, the homelands of the Lenape people.

> The Lenape people are the original inhabitants of Delaware, New Jersey, Eastern Pennsylvania, and Southern New York. For over 10,000 years they have been the caretakers of these lands and of The River of Human Beings, more commonly known as the Delaware River. The Lenape were the first tribe to sign a treaty with the United States and the first tribe to have land set aside for them in New Jersey. Over a period of 250 years, many Lenape people were removed and dispersed throughout the country. Some took refuge with other tribes. A large number of Lenape families remained in the homelands and continue the traditions of their ancestors up to our present day. Today the Lenape people from all over Turtle Island (North America) are revitalizing their communities. (See Lenape Nation of Pennsylvania, https://www.lenape-nation.org)

3. Claudia Vargas, "As Hundreds of Eastwick Homes Sink and Become Unsafe, Residents Look for Help," *Philadelphia Inquirer*, June 20, 2016, https://www.inquirer.com/philly/news/20160620_As_hundreds_of_Eastwick_homes_sink_and_become_unsafe__residents_look_for_help.html; "Harvey-Level Damage Probably Won't Happen in Philadelphia."

4. When I refer to environmental racism, I am drawing on Robert D. Bullard's work that posits it as "any policy, practice, or directive that differentially affects or disadvantages (whether intended or unintended) individuals, groups, or communities based on race or color. It also includes exclusionary and restrictive practices that limit participation by people of color in decisionmaking." Bullard, "Threat of Environmental Racism," 23–26, 55–56. When I refer to different forms of colonialism, I am drawing off of Eve Tuck and K. Wayne Yang's explanation of external colonialism, internal colonialism, and settler colonialism in "Decolonization Is Not a Metaphor." For the purposes of this work, I will be exploring the relations between weather science, data, and settler colonialism, which "operates through internal/external colonial modes simultaneously because there is no spatial separation between metropole and colony" (Tuck and Yang, "Decolonization Is Not a Metaphor," 5); National Resources Defense Council, "Fighting for Safe Drinking Water in Flint."

5. Yusoff, "Anthropogenesis," 9; Tuck and Yang, "Decolonization Is Not a Metaphor," 5.

6. Davis and Todd, "On the Importance of a Date," 764.

7. Kyle Whyte defines settler colonialism—the primary activity that settlers participate in—as "social processes in which at least one society seeks to move permanently onto the terrestrial, aquatic, and aerial places lived in by one or more other societies who already derive economic vitality, cultural flourishing, and political self-determination from the relationships they have established with the plants, animals, physical entities, and ecosystems of those places" (Whyte, "Dakota Access Pipeline," 323–24). Settler colonialism is "structure rather than an event," Patrick Wolfe elaborates, continuing that settler colonialism "destroys to replace" and holds access to land, or "territoriality," as its "specific, irreducible element" (Wolfe, "Settler Colonialism," 388). Tuck and Yang contribute that the settler seeks to make "a new 'home' and that home is rooted in a homesteading worldview where the wild land and wild people were made for his benefit" (Tuck and Yang, "Decolonization Is Not a Metaphor," 6).

8. Voyles, *Settler Sea*, 6.

9. Davis and Todd, "On the Importance of a Date," 764.

10. Environmental Document Access and Display System, Version 2 (EV2) ("East India Company Logbooks").

11. EV2 ("Dutch Log Books").

12. Military Forts and Voluntary Observers Database, accessed June 15, 2022, https://mrcc.purdue.edu/data_serv/cdmp/cdmp.jsp.

13. EV2 ("Iraq Upper Air Data").

14. Perry, "In Bondage When Cold Was King," 24, 32.

15. Pawley, *Nature of the Future*, 5, 18, 198.

16. Lucier, *Scientists and Swindlers*.

17. Valenčius, *Health of the Country*, 143.

18. Anderson, *Predicting the Weather*; Hardy, "Meteorology as Nationalism"; Beattie, O'Gorman, and Henry, *Climate, Science, and Colonization Histories*.

19. Golinski, *British Weather and the Climate of Enlightenment*.

20. Wolfe, "Settler Colonialism," 393.

21. Tuck and McKenzie, "Relational Validity," 635.

22. There is a rich literature within the social sciences on relationality and research, as well as being in relation to the nonhuman world. In "Relational Validity and the 'Where' of Inquiry: Place and Land in Qualitative Research," Eve Tuck and Marcia McKenzie conceptualize "critical place inquiry," which is "a set of concepts, practices, and theories which move beyond understandings of place as a neutral backdrop, or as a bounded and antiquated concept, or as only a physical landscape." Critical place inquiry centers "relational validity," a concept that "prioritizes the reality that human life is connected to and dependent on other species and the land" (636). In "Being in Relation," Kim TallBear writes that nonhumans "have their own life ways, they have a social world, they have social practices, they transmit knowledge. They have their

own intimate relationships among themselves, and with some humans" (60). While "being in relation" is different from "relational validity," both works express relation as a way of existing in and understanding the world. These formations help ground what I mean by relation, or lack of relation, across this book. Whyte, "Indigenous Climate Change Studies"; Liboiron, *Pollution Is Colonialism*.

23. Styres, Haig-Brown, and Blimkie, "Towards a Pedagogy of Land," 38.

24. Liboiron, *Pollution Is Colonialism*, 7n19.

25. Liboiron, *Pollution Is Colonialism*, 7n19.

26. Valenčius, *Health of the Country*, 4–5, 18.

27. Kimmerer, *Braiding Sweetgrass*.

28. This argument would not be possible without the work of many scholars who have argued for and elaborated on the link between either pollution and colonialism, or climate change and colonialism. In *Pollution Is Colonialism*, Max Liboiron "redefine[s] pollution as central to, rather than a byproduct of, colonialism" (36). Kyle Whyte has explored "climate change as intensified colonialism," whereby "Indigenous studies . . . seek to understand vulnerability to climate change as an intensification colonialism" ("Indigenous Climate Change Studies," 3). Both works have helped me conceptualize fundamental relationships between colonialism, climate change, Western science, and harm. See also Heglar, "Climate Change Isn't the First Existential Threat."

29. Pierotti and Wildcat, "Traditional Ecological Knowledge."

30. Smith and Sharp, "Indigenous Climate Knowledges," 468.

31. See Pierotti and Wildcat, "Traditional Ecological Knowledge"; Kimmerer, "Weaving Traditional Ecological Knowledge."

32. Kouri and Skott-Myhre, "Catastrophe," 281.

33. Mitchell and Chaudhury, "Worlding beyond 'the' 'End' of 'the World,'" 311.

34. Mitchell and Chaudhury, "Worlding beyond 'the' 'End' of 'the World,'" 311.

35. King, *Black Shoals*, 68–69.

36. King, *Black Shoals*, 65.

37. King, *Black Shoals*, 62.

38. Yusoff, "Anthropogenesis," 13.

39. This thinking falls in line with the work that Davis and Todd do to problematize Paul Crutzen and Eugene Stoermer's Anthropocene discourse in "On the Importance of a Date," 768.

40. Appadurai, *Modernity at Large*, 116.

41. Smithsonian Institution, *Annual Report of the Board of Regents of the Smithsonian Institution* (1857), 28.

42. "The Ramaytush Ohlone: The Original Peoples of the San Francisco Peninsula." For further information, see the Association of Ramaytush Ohlone website, https://www.ramaytush.org/ramaytush-ohlone.html (accessed November 12, 2020).

43. Smithsonian Institution, *Annual Report of the Board of Regents of the Smithsonian Institution* (1852), 108.

44. Reingold, "Professionalization of Science"; Nebeker, *Calculating the Weather*; Mergen, *Weather Matters*; Harper, *Weather by the Numbers*; Fleming, *Meteorology in America*; Edwards, *Vast Machine*.

45. Tuck and Yang, "Decolonization Is Not a Metaphor," 7.

46. On the history of settler colonialism and land, see Wolfe, "Settler Colonialism," 393.

47. Smithers, *Science, Sexuality, and Race*.

48. Harris, "Whiteness as Property," 1715–18. See also Smithers, *Science, Sexuality, and Race*.

49. Ahmed, "Phenomenology of Whiteness," 153–54.

50. See Wilm, "Poor, White, and Useful," 212–13.

51. Ahmed, "Phenomenology of Whiteness," 158.

52. On property ownership and race in the middle to late nineteenth century, see Finney, "Birth of a Nation."

53. Climate Data Records, National Centers for Environmental Information, accessed January 2023, https://www.ncei.noaa.gov/products/climate-data-records.

54. Hitt, "Mauna Kea Protests Reveal Problems."

55. Hicks, *Programmed Inequality*.

56. On erasure and persistence, see O'Brien, *Dispossession by Degrees*.

57. Dunaway, *Seeing Green*; Howe, *Behind the Curve*.

58. Rosenberg, "Data before Fact"; Ribes and Jackson, "Data Bite Man."

59. D'Ignazio and Klein, *Data Feminism*.

60. Gitelman, *"Raw Data" Is an Oxymoron*; Gitelman, *Always Already New*.

61. Kitchin, *Data Revolution*.

62. Pine and Liboiron, "Politics of Measurement and Action," 3147–48.

63. Rosenberg, "Data before Fact," 32.

64. Daston and Lunbeck, *Histories of Scientific Observation*; Fleming, *Meteorology in America*.

65. Harper, *Weather by the Numbers*; Howe, *Behind the Curve*.

66. Tuck and Gaztambide-Fernández, "Curriculum, Replacement, and Settler Futurity," 72–89; Tuck and Yang, "Decolonization Is Not a Metaphor."

67. On Land as resource, see Max Liboiron's chapter "Land, Nature, Resource, Property," in *Pollution Is Colonialism*. "Resources refer to unidirectional relations where aspects of land are useful to particular (here settler and colonial) ends" (62).

68. Solnit, *Faraway Nearby*, 3.

69. On sense of place, see Worster, *Dust Bowl*, 4, 167. On how data gets "made global," see Edwards, *Vast Machine*, xvi and chap. 10, "Making Data Global."

70. Blum, *News at the Ends of the Earth*.

CHAPTER 1. **Dreaming Data**

1. As the Cayuga Nation outlines,

> The people of the Cayuga Nation have called the land surrounding Cayuga Lake their homeland for hundreds of years. . . . All was stable until the Revolutionary War. Although the Cayuga Nation remained neutral, it

became the target of U.S. military attacks. Cayuga villages were destroyed and its orchards burned during the campaigns of General Sullivan and Colonel Butler. The Cayugas were forced from their homeland and the land was dispersed in parcels to American soldiers. In November of 1794 it appeared that the wrongful taking of Cayuga land would be made right. The Treaty of Canandaigua was signed between the Sachems of the Confederacy Nations and the United States of America. This Treaty affirmed the Cayuga Nation's rightful reservation as 64,015 acres of sovereign land. Unfortunately, the Treaty was ignored by New York. The Cayuga homeland was not returned to its owners. For the next 250 years the Cayuga Nation pursued its land claim against New York State. In the early 21st century we made the decision to take affirmative action. The Cayuga Nation decided to start reacquiring its land by simply purchasing it. (Cayuga Nation: "People of the Great Swamp")

More information about the Cayuga Nation can be found at https://cayuganation-nsn.gov.

2. *Reports of Meteorological Observations Made at Academies.*

3. *Reports of Meteorological Observations Made at Academies.*

4. Jones and Franklin Institute, *Journal of the Franklin Institute* 19:8.

5. Daston, "Super-Vision"; Anderson, *Predicting the Weather*, 237. Within the United States, the navy had begun collecting weather data in 1814 under an order by then surgeon general James Tilton. While they began with prose reports—nonnumerical diaries and journals—by the late 1840s, they had moved on to the newly standardized form, the weather blank. See US Army Medical Department, "Regulations from the Medical Department," 227–28.

6. Golinski, *British Weather and the Climate of Enlightenment.*

7. On standardization, see Fleming, *Meteorology in America*, 92.

8. Mahony, "For an Empire of 'All Types of Climate,'" 29–39. See also Golinski, *British Weather and the Climate of Enlightenment*; Jankovic, *Reading the Skies.*

9. Cullen and Geros, "Constructing the Monsoon," 4.

10. Heymann, "Evolution of Climate Ideas and Knowledge," 588, 593.

11. On internal colonialism, see Tuck and Yang, "Decolonization Is Not a Metaphor," 18.

12. Simmons, "Settler Atmospherics."

13. Gitelman, *Paper Knowledge*. In addition to Gitelman, Daston and Rusnock have attended in particular to the table as an "epistemic genre"; see Daston, "Super-Vision," 194; Rusnock, "Correspondence Networks and the Royal Society," 155–69.

14. Beck and Henry, *Abstract of the Returns of Meteorological Observations.*

15. In the United States, enthusiasts produced lexical weather diaries up until the early nineteenth century. Erratic efforts at measuring weather and placing those measurements in tabular form predate 1825. The Medical Department of the army, for example, began keeping regular "diaries" of the weather in 1814, and these diaries generally followed tabular form. In a general order from May of

that year, James Tilton directed hospital workers, including surgeons, to "keep a diary of the weather . . . a book in which shall be registered all the reports transmitted by him." The diaries kept by medical surgeons were transmitted to the Surgeon General's Office from 1819 up until 1843. In addition, the Land Office tried to keep quantified weather records beginning in 1817 under commissioner Josiah Meigs. The effort was short-lived, though in 1821 at least twelve offices reported handwritten tables based off rudimentary sets of instruments Meigs procured without financial aid from Congress. See US Army Medical Department, "Regulations from the Medical Department," 227–28.

16. Dunbar-Ortiz, *Indigenous Peoples' History of the United States*, 76–77.

17. For more information on the Haudenosaunee, see https://www.haudenosauneeconfederacy.com.

18. Hauptman, *Conspiracy of Interests*, 9.

19. New York (State) and Hough, *Proceedings of the Commissioners of Indian Affairs*, 434.

20. See Hauptman, *Conspiracy of Interests*; Hauptman and McLester, *Oneida Indian Journey*.

21. "Minutes, Academy Trustees," *Trustees Minutes* in *Papers of Joseph Henry*, 1:105. Hereafter this multivolume source is cited as *JHP*, followed by the volume and page numbers. Full details, including all author names and the titles of the individual volumes, are provided in the bibliography. See also Fleming, *Meteorology in America*, 19.

22. De Witt presented a paper, "Functions of the Moon as deduced from the total Eclipse of the Sun on the 16th of June 1806," at the Albany Institute on March 2, 1825, one day after meeting with the Board of Regents to discuss the New York meteorological network.

23. University of the State of New York, *Instructions from the Regents*.

24. The seventeenth- and eighteenth-century European weather table—exemplified by diarists and observers such as John Locke, James Jurin, and Louis Cotte—was structured in a fashion similar to what De Witt, Lansing, and Greig designed. The European table was structured around columns and rows: the columns designated different weather phenomena such as temperature, barometric pressure, wind, rain, and "weather," while the rows organized time in relation to these categories. Throughout the seventeenth and early eighteenth centuries, weather observers, or "weather watchers," as Lorraine Daston calls them, were drawn to the table because it could represent "rule-like relationships between the elements of the columns and rows," as well as "draw together phenomena scattered across space as well as time." Weather watchers hoped to achieve a kind of "super-vision" through the time and space of weather. Though the table dominated as the "preferred visual tool" in the late seventeenth and early eighteenth centuries, nonstandard visual conventions hindered "tabular literacy," making seventeenth-century "super-vision" impossible. See Daston, "Super-Vision," 194, 210.

25. Many early weather data collection projects, including those in Europe, circulated guidelines for data recording in the form of sample tables. In 1723,

Jacob Jurin published "A Proposal for Joint Observations on the Weather," which included a sample weather table in Latin. See Jurin, "Invitatio ad observationes." Meteorological projects in Maryland and Philadelphia used this tabular style as well.

26. "Instructions for Meteorological Reports, 1825," 48.

27. *Reports of Meteorological Observations Made at Academies.*

28. *Reports of Meteorological Observations Made at Academies.*

29. University of the State of New York, *Instructions from the Regents* (1836), 18.

30. University of the State of New York, *Instructions from the Regents* (1836), 18–19.

31. University of the State of New York, *Instructions from the Regents* (1836), 37–38.

32. University of the State of New York, *Instructions from the Regents* (1836), 53.

33. University of the State of New York, *Annual Report of the Regents* (1837), 220.

34. University of the State of New York, *Annual Report of the Regents* (1837), 222.

35. University of the State of New York, *Fifty-Third Annual Report of the Regents* (1840).

36. "Galvanism," September 6, 1833, *JHP*, 2:90–96.

37. "Galvanism," September 6, 1833, *JHP*, 2:90–96.

38. Letter from Peter Bullions to Joseph Henry, December 19, 1833, *JHP*, 2:138–41.

39. Letter from Joseph Henry to William Kelly, December 30, 1833, *JHP*, 2:142.

40. As the language of data slides into view, it slips between operating as a "premise" of investigation at the beginning of the century to the "result of investigation" at the end. See Rosenburg, "Data before the Fact."

41. *JHP*, 1:435–37. Henry wrote this letter in response to Maclean's note of June 18, in which Maclean made Henry aware of the soon-to-be vacant chair of the Department of Natural Philosophy at Princeton (a position he would take later that year).

42. There is also a religious tone in Henry's letter to Beck; the aurora as a "sign in the heavens that Mrs. Whipple should have been hung with Jessy Strang." As the editors of *JHP* note, Whipple and Strang were accused of the murder of John Whipple, but only Strang was hung for the crime. Beck returned Henry's letter, writing that the northern lights have all but disappeared, and "I look upon this as convincing proof that they must have had an intimate connexion with the affair of Strang & Mrs. Whipple. Your electro-magnetic theory therefore must be given up." Letter from T. Romeyn Beck to Joseph Henry, October 7, 1827, *JHP*, 1:200; "Excerpt, Minutes, Albany Academy," April 25, 1831, *JHP*, 1:338.

43. Letter from Joseph Henry to Parker Cleveland, February 14, 1832, *JHP*, 1:402–3. Henry was gathering observations about the aurora because he had hoped to, and indeed would, publish an account of the meteorological phenomenon in *Sillman's Journal* under the title "On a Disturbance of the Earth's Magnetism, in Connexion with the Appearance of an Aurora Borealis, as Observed at Albany, April 19th, 1831."

44. Henry, *Scientific Writings*, 72.

45. Letter to from Joseph Henry to Alexander Dallas Bache, March 9–27, 1839. It was dated in such a way because it was started on March 9 and not finished until March 27. *JHP*, 4:186–92.

46. At least three additional philosophical societies and academic institutions were trying to build regional networks to measure weather and publish their results in tabular form. In 1838 the American Philosophical Society funded a magnetic and meteorological observation station at Girard College in Philadelphia, and results from these early years of observation, headed by Alexander Dallas Bache, were published by Gale and Seaton Printers in 1847 Washington, DC. Note also that the state of New York's table design also overlapped with tabular design at the Franklin Institute, American Philosophical Society, in coordination with Girard College.

47. Weather "prognostics" had been published in the *Journal of the Franklin Institute*, for example, since 1828.

48. Espy, "On the Importance of Hygrometric Observations," 222.

49. See Hauptman, *Conspiracy of Interests*; Haudenosaunee Confederacy, "Land Rights and Treaties."

50. Barker, "Territory as Analytic," 26; Dunbar-Ortiz, *Indigenous Peoples' History of the United States*, 111.

51. Espy, "On the Importance of Hygrometric Observations," 144.

52. Jones and Franklin Institute, *Journal of the Franklin Institute* 18:383.

53. Jones and Franklin Institute, *Journal of the Franklin Institute* 19:7–8.

54. Jones and Franklin Institute, *Journal of the Franklin Institute* 19:7–8.

55. Jones and Franklin Institute, *Journal of the Franklin Institute* 19:13.

56. Jones and Franklin Institute, *Journal of the Franklin Institute* 19:17.

57. Jones and Franklin Institute, *Journal of the Franklin Institute* 19:13.

58. Jones and Franklin Institute, *Journal of the Franklin Institute* 19:19.

59. Jones and Franklin Institute, *Journal of the Franklin Institute* 19:17–18.

60. Jones and Franklin Institute, *Journal of the Franklin Institute* 19:20.

61. Jones and Franklin Institute, *Journal of the Franklin Institute* 19:144.

62. Jones and Franklin Institute, *Journal of the Franklin Institute* 19:194.

63. Espy, *First Report on Meteorology*, 2.

64. Espy, "To the Honorable."

65. For a detailed account of this controversy, see Fleming, *Meteorology in America*, 23–54.

66. For a detailed account of Espy's lyceum participation and Espy and Henry's European tour, see Fleming, *Meteorology in America*, 43, 45.

67. In this letter, Forbes also wrote that Henry would "probably be interested although you may not be inclined to adopt all his views." Henry's alliances were tested by the storm controversy, and Forbes's suggestion here is evidence of that. Letter from James Forbes to Joseph Henry, June 3, 1845, *JHP*, 6:289–91.

68. Espy submitted these reports up until 1857; see *JHP*, 7:xxvii for notes regarding Henry's funding of Espy's work at the Surgeon General's Office.

69. Espy, *First Report on Meteorology*, 1.

70. Jones and Franklin Institute, *Journal of the Franklin Institute* 7:222.

71. See Espy, *Second Report on Meteorology*.

72. See Espy, "Circular to the Friends of Science," 84. The circular carries the date of October 9, 1843.

73. Published in 1841, a year before Espy began his work with the Surgeon General's Office, Espy's *Philosophy of Storms* notes that "data are not yet sufficient" to determine the number of storms in particular scenarios, as well as data being "imperfect." Espy sought more data as well as standardized data, an effort that would continue in his work through the Surgeon General's Office. See Espy, *Philosophy of Storms*, 242, 254.

74. Coffin, *Meteorological Register and Scientific Journal*, 1.

75. Appadurai, *Modernity at Large*, 116. After discarding centuries of Indigenous knowledge on weather and climate through colonization, weather observers in the United States used tables, data, and the "language of numbers" to secure colonial space, 116.

76. Clyde, *Life of James H. Coffin*; see also *JHP*, 5:266.

77. Letter from Joseph Henry to James Coffin, September 9, 1842, *JHP*, 5, 267. Henry explains in his letter that though he "commenced" a series of observations at Princeton, while he was abroad, they were "discontinued." Other than Henry's attempt to establish a series, a register of the thermometer was kept by Dr. Hodge of the Theological Seminary.

78. Coffin had been working on mapping winds since 1836, first at Dartmouth and then at Ogdensburg. His work on winds had been published under the general rubric of meteorological observations in the New York State Regents' reports in 1838 and 1840 (see Coffin, Voeïkov, and Coffin, *Winds of the Globe*, xvi). It is unclear whether Henry replied to Coffin's July request.

79. Letter from James Coffin to Joseph Henry, December 21, 1846, *JHP*, 6:603.

80. As editors of the *JHP* note, Coffin's meteorological efforts depended on "networks of observers sending data to a central point for analysis." *JHP*, 6:603n1.

81. Smithsonian Institution, *Third Annual Report of the Board of Regents* (1853), 161.

82. Henry and Espy, "Circular from the Smithsonian Institution."

83. See Record Unit 27.3 at the National Archives and Record Unit 60 at the Smithsonian Institution Archives.

84. Observers were located in present-day Illinois, Kentucky, Massachusetts, Maine, New Jersey, New York, North Carolina, Pennsylvania, Virginia, and Wisconsin.

85. Smithsonian Institution Archives, Record Unit 60.

86. American Association for the Advancement of Science, *Proceedings from the Meeting*, 62.

87. Henry, "Programme of Organization."

88. Espy, "First Report on Meteorology," 1.

89. Smithsonian Institution, *Annual Report of the Board of Regents* (1849), 15.

90. Letter from Joseph Henry to Harriet Henry, September 12, 1859, *JHP*, 10:106.

91. Letter from Joseph Henry to Daniel William Fiske, March 16, 1960, *JHP*, 10:141; letter from Joseph Henry to John Henry Lefroy, October 24, 1860, *JHP*, 10:169.

92. Takarabe, "Smithsonian Meteorological Project," 6.

93. "Journal of the Proceedings of the Regents of the Smithsonian Institution," 1848, 15.

94. Smithsonian Institution, *Annual Report of the Board of Regents* (1865), 43.

95. Smithsonian Institution, *Annual Report of the Board of Regents* (1852), 108.

96. Smithsonian Institution, *Annual Report of the Board of Regents* (1854), 25.

97. Smithsonian Institution, *Annual Report of the Board of Regents* (1855), 28.

98. Smithsonian Institution, *Tenth Annual Report of the Board of Regents* (1856), 263.

99. Bishop et al., *Results of Meteorological Observations*.

100. Bishop et al., *Results of Meteorological Observations*, iv.

101. Bishop et al., *Results of Meteorological Observations*, vii.

102. Heymann, "Evolution of Climate Ideas and Knowledge," 587.

103. Liboiron, *Pollution Is Colonialism*, 46.

104. Liboiron, *Pollution Is Colonialism*, 48n33.

CHAPTER 2. **Gendering Data**

Grateful acknowledgment is made to the *Journal of Women's History*, which published portions of this chapter. Sara J. Grossman, "Gendering Nineteenth-Century Data: The Women of the Smithsonian Meteorological Project," 85–109.

1. UmoNhoN means "against the current." For more information about the UmoNhoN, see the Omaha Tribal Historical Research Project, http://jackalopearts.org/othrp.htm.

In 1847, the Smithsonian Institution began the first national volunteer weather data collection project, which utilized volunteers from across the county (not simply from the eastern region) to perform daily weather measurements. Measurements were sent to the Smithsonian monthly and eventually led to the formation of the National Weather Service. See Fleming, *Meteorology in America*.

2. See Cession 315 in Smithsonian Institution Bureau of American Ethnology, *Annual Report of the Bureau of American Ethnology*, 791. For thirteen years—the Bowens claimed this land as their own, despite its being the ancestral lands of the Omaha—and in 1870 they filed for property rights under the Homestead Act. U.S., Homestead Records, 1863–1908, Land Entry Case Files: Homestead Final Certificates, Record Group 49, Records of the Bureau of Land Management.

3. "Face of the sky" was a data recording technique that relied on sensory observation to denote cloud cover, the force of the wind, rain, and snow.

4. Bowen's status as an observer is well documented in the Smithsonian Institution's *Annual Report of the Board of Regents* from 1850 to 1864.

5. Light, "When Computers Were Women," 483; Hicks, *Programmed Inequality*.

6. D'Ignazio and Klein, "Bring Back the Bodies" and "Show Your Work," in *Data Feminism*.

7. Loukissas, "A Place for Big Data," 4.

8. On the emergence of professional science, see Daniels's periodization in *The Process of Professionalization*. See also Reingold, *Science in Nineteenth-Century America*, 128; Kohlstedt, "In from the Periphery," 81–96; Swanson, "Rubbing Elbows and Blowing Smoke," 40–61.

9. See Hoffert, "Female Self-Making," 34–59.

10. See Hoffert, "Female Self-Making."

11. Abbate, *Recoding Gender*, 4.

12. Tolley, *Science Education of American Girls*, 44.

13. Baumgartner, *In Pursuit of Knowledge*, 111.

14. Baumgartner, *In Pursuit of Knowledge*, 5.

15. Fixico, *Bureau of Indian Affairs*, 70.

16. Smithsonian Institution, *Annual Report of the Board of Regents* (1873), 76.

17. Mihesuah, *Cultivating the Rosebuds*, 55, 61, 89.

18. Fleming, *Meteorology in America*, 126.

19. "Sketch of Professor Coffin."

20. Kohlstedt, "In from the Periphery," 81; Lee, *Hidden Figures*.

21. On the production of computational skill as a gendered trait in the history of computing, see Abbate, *Recoding Gender*, 40.

22. Hoffert, "Female Self-Making," 50.

23. Although this chapter does not explore the ways that weather and environmental knowledge were leveraged by Black communities in the middle and late nineteenth century in ways that were distinct from white elites, several historians have documented this vital history. See Roane, "Plotting the Black Commons," 239–66; Perry, "In Bondage When Cold Was King," 23–36.

24. On the relationship between settler colonialism and heteropatriarchy, see Arvin, Tuck, and Morrill, "Decolonizing Feminism," 8–34.

25. P. A. Chadbourne worked for the Theological Institute of Connecticut. Record Group 29, *Seventh Census of the United States, 1850*.

26. Smithsonian Institution Archives, Record Unit 60 (hereafter RU 60), Smithsonian Institution, Meteorological Project, Records.

27. Fleming, *Meteorology in America*, 81.

28. RU 60.

29. Even though data was everywhere—from newly discovered flora to fauna—amateurs "were confronted with exceptionally challenging problems" of categorization and classification. See Reingold, "The Professionalization of Science," 45.

30. RU 60.

31. RU 60.

32. RU 60.

33. RU 60.

34. RU 60.

35. Smithsonian Institution, *Annual Report of the Board of Regents* (1855), 245.

36. Smithsonian Institution, *Annual Report of the Board of Regents* (1855).

37. Smithsonian Institution, *Annual Report of the Board of Regents* (1849), 30.

38. Record Group 29, *Seventh Census of the United States, 1850*.

39. "Researches as to the Phenomena of American Storms," *Semi-Weekly Eagle*, March 5, 1849, 2.

40. "Circular on Meteorology," November 1, 1848, RU 60.

41. Smithsonian Institution, *Annual Report of the Board of Regents* (1852), 17.

42. Smithsonian Institution, *Annual Report of the Board of Regents* (1856), 26.

43. Data collected by author, unpublished.

44. Each observer noted here provided letters and data sheets to the Smithsonian Institution. Some of these letters were saved by the various caretakers of the Smithsonian Meteorological Project archival holdings and can be found in RU 60, as well as in the National Archives, Record Group 27.

45. Letter from Susannah Spencer to Smithsonian Institution, March 13, 1853, RU 60.

46. Letter from Maria Mitchell to Joseph Henry, September 10, 1855, RU 60. The Smithsonian noted that another letter was received on September 19, 1855, but this has not been found. See Smithsonian Institution, *Annual Report of the Board of Regents* (1869), 75, for an observer list where only Mitchell's father's name appears. Henry had known Maria Mitchell and her father since 1848, when Henry agreed to publish Mitchell's paper outlining her comet discovery in the second volume of *Smithsonian Contributions to Knowledge* (see JHP, 7:344nn3, 4). Though Henry supported the publication of Mitchell's discovery, male members of the review board objected, remarking that the "discovery of a New Comet by an American Lady" would not serve to "diffuse in the aggregate a greater amount of knowledge among men, that is *all* men" (letter dated June 3, 1849, JHP, 7:544).

47. See Kohlstedt, "In from the Periphery," 87.

48. Letter from Sara Thomas to Professor Henry, 1859, RU 60.

49. "Meteorological Instruments Loaned, 1850–1870," 1 volume, arranged by states with an alphabetical index to persons, RU 60.

50. While Bowen's original letter and drawing have been lost, a secondary note (made by the Smithsonian Institution) survives in RU 60 today.

51. Record Group 27, Records of the Weather Bureau (hereafter cited as RG 27).

52. Record Group 29, *Seventh Census of the United States, 1850*.

53. Letter dated February 21, 1849, RG 27.

54. Smithsonian Institution, *Annual Report of the Board of Regents* (1869), 75.

55. Letter dated February 4, 1850, RG 27.

56. Letter dated September 30, 1851, RG 27.

57. Letter dated September 2, 1851, RG 27.

58. Letter dated January 10, 1852, RG 27.

59. Smithsonian Institution, *Annual Report of the Board of Regents* (1869), 75.

60. Smithsonian Institution, *Annual Report of the Board of Regents* (1873), 65.

61. See Fleming, *Meteorology in America*; see also JHP, 9:xx. Henry dismissed Blodget because he had been using incoming volunteer data for his own scien-

tific papers. Henry argued that all incoming data was the property of the institution and the volunteers who created it and not intended for Blodget's personal use. After notifying Blodget that his services were no longer needed by the institution, Henry had the locks on Blodget's door changed while Blodget was out to dinner on October 11, 1854.

62. Penny, *Employments of Women*, 16.

63. On women gaining admittance through male sponsorship and mentorship, see Hoffert, "Female Self-Making," 49.

64. Record Group 29, *Seventh Census of the United States, 1850*.

65. See RU 60; National Archives at College Park, RG 27.

66. Letter from Harriet Baer to Smithsonian Institution, RU 60.

67. See letters from September 1858 to early 1859, RU 60, especially those of J. D. Parker of Maine, who was sent a brass rain gauge in 1859, and Franklin Fairbanks, of Vermont, who corresponded with Henry about an "Equinoctial Storm" in September of 1858.

68. RU 60.

69. *JHP*, 8:xxii.

70. "Smithsonian Institution Board of Regents, Minutes, 1854," Smithsonian Institution Archives, Record Unit 1, Board of Regents vol. 1.

71. *JHP*, 10:xx.

72. *JHP*, 10:xx.

73. See RG 27.

74. On the difficulty of recovering "amateur" labor, see Gotkin, "When Computers Were Amateur," 4–14.

75. Letter from James Coffin to Joseph Henry, February 25, 1859, Smithsonian Institution Archives, Record Unit 7060.

76. Record Group 29, *Eighth Census of the United States, 1860*.

77. Letter from James Coffin to Joseph Henry, February 25, 1859.

78. See Henry's specification of laborers as "female" in Smithsonian Institution, *Annual Report of the Board of Regents* (1857), 28.

79. See Henry's description of meteorological data labor in "Report of the Secretary of the Smithsonian Institution," in Smithsonian Institution, *Annual Report of the Board of Regents* (1856), 27–28.

80. Bishop et al., *Results of Meteorological Observations*.

81. Swanson, "Rubbing Elbows and Blowing Smoke," 48; Kohlstedt, "In from the Periphery," 89.

82. Smithsonian Institution Archives, Joseph Henry Collection, Record Group 7001.

83. Smithsonian Institution, *Annual Report of the Board of Regents* (1856), 27.

84. For the history of data overloads, see Rosenberg, "Early Modern Information Overload," 1–9.

85. Letter from James Coffin to Joseph Henry, March 16, 1855, James H. Coffin Papers, 1829–1911.

86. Coffin refers to the superintendent as male in his letter.

87. This portion of Coffin's letter is almost unreadable. I have tried to the best of my ability to reconstruct this sentence faithfully.

88. James Coffin to Joseph Henry, July 12, 1855, Smithsonian Institution Archives, Record Unit 7001, Joseph Henry Collection.

89. Smithsonian Institution, *Annual Report of the Board of Regents* (1857), 28.

90. There is additional evidence that James Coffin employed women in the work of reduction. In *The Life of J. H. Coffin*, Coffin's son notes that his father "engaged the services" of "a considerable number of ladies in Easton" and that "the wages of these computers were paid by a small appropriation from Congress." John Cunningham Clyde, *Life of James H. Coffin*, 65.

91. Reduction was intimately tied to computing at the time. My work takes a wide view of reduction as a computing practice that encompasses nineteenth-century human computing, accounting for people who "reduced or analyzed data using mechanical calculators" as Paul Ceruzzi notes, as well as those who, according to Peggy Kidwell, performed reduction through the use of "mathematical tables and hand reckoning." See Ceruzzi, "When Computers Were Human," 237; Kidwell, "American Scientists and Calculating Machines," 31–40; Light, "When Computers Were Women."

92. See Smithsonian Institution, *Annual Report of the Board of Regents* (1857), 27–28.

93. This is my own math based on Henry's remarks on yearly totals.

94. On the interior cover of Lafayette College's *Results of Meteorological Observations, 1854–1859*—likely given to Coffin by Mason and Henry—is inscribed, "Professor James H. Coffin, LL.D. was for seven years engaged in preparing this volume, 1856–1863. In the compilation, he was aided by twenty students of Lafayette, and fourteen teachers of Easton schools." James H. Coffin Papers, 1829–1911. See also Bishop et al., *Results of Meteorological Observations*.

95. Letter from James Coffin to Joseph Henry, March 16, 1855.

96. In a letter from Coffin to Henry dated March 1, 1855, Coffin proposes the rate of twenty-five cents per hour. This rate is written into the contract that Henry and Coffin sign on June 15, 1855. Coffin raises the possibility of employing women for this reduction work in a letter later that month, March 16, 1855, and Henry confirms this in the institution report of 1856. See also JHP, 9:256–59.

97. See letter from Charles Mason to Joseph Henry, 1855, JHP.

98. Sellers, *Commissioner Charles Mason and Clara Barton*, 814. Sellers notes that as early as 1848, five temporary clerks were employed on a part-time basis at the Patent Office. Compensation for the year ranged from "$44.19 to $174.11" (811).

99. Epler, *Life of Clara Barton*, 24.

100. Pryor, *Clara Barton*, 56.

101. Pryor, *Clara Barton*, 56.

102. Sellers, *Commissioner Charles Mason and Clara Barton*, 813.

103. Sellers, *Commissioner Charles Mason and Clara Barton*, 813.

104. "Miscellaneous Letters. Files of the Department of the Interior, Archives

of the United States," quoted in Sellers, *Commissioner Charles Mason and Clara Barton*, 814–16.

105. "Outgoing Correspondence of the Secretary of the Interior. Patents and Miscellaneous Division. Vol. 1," quoted in Sellers, *Commissioner Charles Mason and Clara Barton*, 813.

106. Swanson, "Rubbing Elbows and Blowing Smoke," 59. In an 1865 letter to Henry Wilson, Clara Barton reflected on her time at the Patent Office: "You are aware of the fact that I was, among my sex a pioneer in that time having been one of the very first women who ever held such position, and met, and endured whatever of jealous bickerings and malignity grew of such evident innovation upon the customary rights of man. . . . I naturally fell into political disfavor and was early informed that being a 'Black Republican' I was 'disloyal,' and no longer worth the patronage of the government. My books were demanded, three hundred and eighty dollars of undrawn salary withheld from me, on settlement, and I returned to New England to wonder in silence if god ruled or reigned." Letter from Barton to Henry Wilson, December 19, 1865, Papers of Clara Barton, reel no. 63.

107. Swanson, "Rubbing Elbows and Blowing Smoke," 49.

108. Swanson, "Rubbing Elbows and Blowing Smoke," 49.

109. While Lorin Blodget published what many historians deem the first account by way of his *Climatology of the United States* (1857), it is important to note that Blodget used Smithsonian volunteer data without crediting the volunteers themselves. Though he referred to the Smithsonian Institution and the geographic distribution of observation locations in his report, he did not credit individual volunteers, nor did he name them. This would not have been a difficult task, as Henry managed to provide, and continued to update, observer names throughout the weather data publications he produced in his tenure as secretary.

110. Kelley, *Learning to Stand and Speak*, 7.

111. Boylan, "Claiming Visibility," 21.

CHAPTER 3. **Data in the Sky**

1. "Ornamental Kites," 24.

2. "Ornamental Kites," 24.

3. Subscription required a permanent address and a payment of two dollars per year. Mott, *History of American Magazines*, 317.

4. "Wild Man of the Mountains," 1; "Thoughts for the Working Men," 1.

5. Chudacoff, *Children at Play*, 55; Golinski, "Care of the Self."

6. Haffner, *View from Above*, 15.

7. Signal Service, *Annual Report of the Signal Service*. See list and explanation of stations.

8. See station histories for Wickenburg, Arizona, and Forte Verde, Arizona, through the Forts project. Though not explored here, Washita, Oklahoma, was also a meteorological fort; see https://mrcc.purdue.edu/FORTS/histories1.jsp. In Wickenburg, Arizona, and Forte Verde, Arizona, Cession 689 was used to dis-

place the Apache and Hopi of their homelands. Ancestors of the White Mountain Apache, Tribe of the Fort Apache Reservation, Arizona, "were able to maintain a portion of [our] homeland as the White Mountain Apache Reservation." The White Mountain Apache Tribe "now consists of approximately 16,000 Tribal members. Many live here on our Tribal lands, but others live and work all over the country and throughout world. The majority of the population lives in and around Whiteriver, the seat of Tribal government, with others residing in the communities of Cibecue, Carrizo, Cedar Creek, Forestdale, Hon-Dah, McNary, East Fork, and Seven Mile." For more on the White Mountain Apache, see the White Mountain Apache Tribe's website, http://www.wmat.us. The Hopi Tribe—a sovereign nation—is now located in northeastern Arizona. The Hopi Tribe official website elaborates that "since time immemorial the Hopi people have lived in Hopitutskwa and have maintained our sacred covenant with Maasaw, the ancient caretaker of the earth, to live as peaceful and humble farmers respectful of the land and its resources." The Hopi reservation "encompasses more than 1.5 million acres, and is made up of 12 villages on three mesas." For more on the Hopi, see the Hopi Tribe's website, https://www.hopi-nsn.gov.

9. Gabrys, *Program Earth*, 5.

10. Anderson, *Imagined Communities*, 47; Loughran, *Republic in Print*.

11. Anderson, *Imagined Communities*, 48.

12. Krause, "Reporting and the Transformations of the Journalistic Field," 93.

13. On frontier culture and resource extraction, see Tsing, *Friction*; Ruiz, *Slow Disturbance*.

14. It is important to remember that white women attempted to stake their claim to data in the mid-nineteenth century, as chapter 2 of this book explores.

15. Letter from Espy to Henry, July 15, 1836, *JHP*, 3:77.

16. Nicollet, *Essay on Meteorological Observations*, 40.

17. "Fatal Accident in Flushing," *New York Times*, March 31, 1856.

18. "General City News," *New York Times*, December 16, 1861; "A Warning to Boys—A Fatal Accident," *New York Times*, March 13, 1862; "Run Over," *New York Times*, February 28, 1861.

19. "Kite Flying Forbidden," *New York Times*, August 14, 1875.

20. *Scientific American* was a self-titled "family newspaper," conveying "more useful intelligence to children and young people, than five times its cost in school instruction." It was available in Boston, Philadelphia, and New York, costing two dollars for a year's subscription; the first issue was published on August 28, 1845.

21. "Lightning Conductors," 277.

22. "The Telegraph," 330.

23. "Japanese Bird Kite," 258; "Danger from Street Telegraph Wires," 278.

24. Maillot performed these experiments in Grenoble. "Gigantic Kite," 310.

25. "Penny Kites," 130.

26. See "Miscellaneous Inventions," 186.

27. Middleton, *Invention of the Meteorological Instruments*, 290–91.

28. McFarland, *Concise History of the U.S. Air Force*, 1. The Seminoles—or *yat'siminoli*, meaning "free people"—"have lived in what is now the Southeastern United States for at least 12,000 years." Broken treaties from the US settler states allowed for the theft of over five million acres of their ancestral land. The Seminole Wars, which took place across the early nineteenth century, led to the murder of many Seminole people, as well as their forced displacement to Oklahoma. Across the nineteenth and twentieth centuries, the Seminole people resisted US occupation. Today, the Seminole people live in Florida and Oklahoma and are part of the Seminole Nation of Oklahoma, the Seminole Tribe of Florida, the Miccosukee Tribe of Indians of Florida, and the unaffiliated Independent or Traditional Seminoles. For more on the Seminole people, see https://www.semtribe .com/stof.

29. Parkinson, "United States Signal Corps Balloons," 190–92.

30. Parkinson, "United States Signal Corps Balloons," 192.

31. Parkinson, "United States Signal Corps Balloons," 202.

32. Whitnah, *History of the United States Weather Bureau*, 102.

33. Captive balloons were studied and used alongside weather kites at Mount Weather, Virginia, between 1904 and 1909. Both were intended to provide the Weather Bureau with a means of upper-air research. In 1909, the Weather Bureau decided to experiment with weather balloons in the West, hoping to send one free balloon into the center of a storm, while another balloon was launched away from the storm. Eight ascents were conducted in 1909 at stations near Omaha, Nebraska, and Indiana with 79,200 feet as the highest altitude reached (Whitnah, *History of the United States Weather Bureau*, 102–4).

34. In 1891 settlers employed kites in the production of "artificial" rain. "Artificial Production of Rain," 145.

35. For Eddy as an "experimenter," see "Aerial Navigation," 114.

36. Fleming, *Meteorology in America*, 149.

37. Fleming, *Meteorology in America*, 147–48.

38. Whitnah, *History of the United States Weather Bureau*, 19.

39. Whitnah, *History of the United States Weather Bureau*, 20; on the relation between the national weather service and a telegraphic surveillance network, see Fleming, "Telegraphing the Weather."

40. Whitnah, *History of the United States Weather Bureau*, 23.

41. Whitnah, *History of the United States Weather Bureau*, 60–61.

42. "Pictures from the Sky: The Crowds in Park Row Photographed from Midair," *New York Times*, November 4, 1896, 16; Moffett, "Scientific Kite-Flying," 379–92.

43. See "W. A. Eddy, Kite Flyer, Dead: One of the First to Experiment with Aerial Photography," *New York Times*, December 27, 1909, 7.

44. As explored later in this chapter, camera-mounted meteorological kites would become important tools for military surveillance.

45. "Observatories in the Sky: Wonderful Results Obtained by Kite-Flying Experiments," *New York Times*, September 4, 1895, 13. It is important to note that

although Eddy's technologies were eventually adopted by the Weather Bureau, he did not work for the bureau when he began his experiments. He was an accountant by day and an amateur tinkerer and weather enthusiast by night, experimenting for the most part on his own and reporting his findings to those who were interested. He published often in newspapers, discussing in basic terms the advances he had made in kite weather measurement and aerial "earth" photography.

46. Conover, *Blue Hill Meteorological Observatory*, 17.

47. Some members of this community had also acted as volunteer observers for the Smithsonian Meteorological Project. Cleveland Abbe, who would become the chief of the Weather Bureau after its 1891 inauguration, served as a volunteer during the Smithsonian project and was a key figure in upper-air exploration through the use of kites.

48. While Rotch and Eddy were both educated beyond the primary level—Rotch having attended MIT and Eddy having taken a prep course at Chicago University—Rotch came from an economically advantaged background. This economic advantage enabled Rotch to open up an observatory, as well as travel abroad in pursuit of the latest science.

49. "W. A. Eddy, Kite Flyer, Dead," 7.

50. Rotch, *Use of Kites to Obtain Meteorological Observations*, 98.

51. White, "Diamonds in the Sky."

52. Fergusson, "Kite Experiments," 323.

53. Fergusson, "Kite Experiments," 323.

54. The "law of kite dynamics" is Eddy's phrase and reflects the nuance with which Eddy approached his work. Langley, and other Smithsonian Institution and government officials, would settle on a more narrowly scientific way to refer to these studies, namely, "experiments in aeronautics." See Langley's 1891 publication "Experiments in Aerodynamics."

55. Smithsonian Institution Archives, Record Unit 31 (hereafter RU 31), Smithsonian Institution, Office of the Secretary, Correspondence.

56. Letter from W. A. Eddy to Samuel Langley, October 19, 1891, RU 31.

57. Letter from W. A. Eddy to Samuel Langley, October 9, 1891, RU 31.

58. Letter from W. A. Eddy to Samuel Langley, October 9, 1891, RU 31.

59. Letter from W. A. Eddy to Samuel Langley, November 9, 1891, RU 31.

60. Letter from W. A. Eddy to Samuel Langley, November 4, 1893, RU 31.

61. Letter from W. A. Eddy to Samuel Langley, November 9, 1891, RU 31.

62. Letter from W. A. Eddy to Samuel Langley, March 29. 1894, RU 31.

63. Letter from W. A. Eddy to Samuel Langley, November 4, 1893, RU 31.

64. Eddy and Clayton, "Eddy Malay Tailless Kite," 169.

65. Letter from W. A. Eddy to Samuel Langley, March 29, 1894, RU 31.

66. Smithsonian Institution, *Annual Report of the Board of Regents* (1894).

67. "W. A. Eddy, Kite Flyer, Dead," 7.

68. "Signaling from Kites," *New York Times*, July 10, 1897.

69. "Flying Machines in America," 13–14.

70. Haffner, *View from Above*, 15.

71. *The Photo-Miniature*, 173.

72. Eddy, *Eddy Family in America*, 417.

73. "For Recording Changes," *New York Times*, June 28, 1896, 25–28. Interestingly, the investigation of the upper air in Washington, DC, and across the nation coincided with moves to "legalize" the air as a building space for skyscrapers in city spaces at the time. See Sullivan, "High-Building Question."

74. The writer frequently discusses the need to redesign instruments for upper-air movement, as many instruments were solely and originally designed for ground-level data collection. Though I do not have the space to discuss this in detail, it would be helpful to locate the various ways these instruments were redesigned and who was performing these modifications.

75. For a comprehensive list of these articles, see the *Monthly Weather Review* online archive housed through the American Meteorological Society website. Alternatively, NOAA's archives hold print copies of the journal from this period. Titles include McAdie, "Fog Studies on Mount Tamalpais," 492–93; Mitchell, "Records by the Kite Corps," 539–40; Eddy, "A Record of Some Kite Experiments," 450–52.

76. Wilson, *McClure's Magazine and the Muckrakers*, 63.

77. Moffett, "Scientific Kite-Flying," 391.

78. Eddy, "Science and Kite Flying," 2333.

79. Marvin, "Weather Bureau Kite," 418–20.

80. Marvin, "Weather Bureau Kite," 418.

81. Marvin, "Weather Bureau Kite," 418.

82. Marvin, "Weather Bureau Kite," 418.

83. In 1901, the *New York Times* published a short piece on advances in wireless telegraphy and weather kites: "William A. Eddy says he is writing to Marconi, inclosing diagrams of an improved tandem-kite system for reaching a height of 3,000 with major Baden-Powell's kites. He thinks this system will enable Marconi to send messages at least 1,000 miles because every increase in height has so far enormously increased the range of wireless telegrams." See "Kite as Aid to Wireless Telegraph," *New York Times*, December 13, 1901, 5.

84. Record Group 27, Records of the Weather Bureau.

85. Recall the wind columns that utilized the Beaufort scale, and the linguistic shorthand for direction.

86. Fergusson, "Exploration of the Air by Means of Kites," 41–42.

87. "William A. Eddy's Kite Experiments: Interesting Facts Developed at the Blue Hill Observatory," *New York Times*, September 5, 1895, 9.

88. Rotch, *Use of Kites to Obtain Meteorological Observations*, 98.

89. For post-1870 weather tables, see Schott, *Tables of Rain and Snow in U.S.*; Schott, *Tables and Results of the Precipitation*; Schott, *Tables of Temperature in U.S.*; Schott, *Temperature Chart of U.S.*

90. United States Weather Bureau, *Special Report of Chief of the Weather Bureau* (1891), 548.

91. United States Weather Bureau, *Special Report of Chief of the Weather Bureau* (1891), 549.

92. United States Weather Bureau, *Special Report of Chief of the Weather Bureau* (1891), 549.

93. United States Weather Bureau, *Special Report of Chief of the Weather Bureau* (1891), 549.

CHAPTER 4. **Data's Edge**

1. United States and US Civil Service Commission, *Official Register of the United States*, 1067.

2. See US Department of Agriculture, Weather Bureau form 1001-Met'l, June 1900, held by NOAA's National Centers for Environmental Information, Center for Weather and Climate, Los Angeles, CA.

3. US Department of Agriculture, Weather Bureau form 1001-Met'l.

4. Throughout his tenure with the Weather Bureau, Franklin had developed some local fame. See Brimmner, *Rain Wizard*.

5. US Department of Agriculture, Weather Bureau form 1001-Met'l.

6. United States, Weather Bureau, *Special Report of Chief* (1891), 579.

7. United States, Weather Bureau, *Special Report of Chief* (1900), 11.

8. Porter, *Rise of Statistical Thinking*.

9. For the standardization of tabular form, see Rusnock, "Correspondence Networks and the Royal Society"; Heymann, "Evolution of Climate Ideas and Knowledge"; Daston, "Super-Vision."

10. Although this chapter addresses Weather Bureau meteorology, the emergence of professional meteorology is an important development in this period. For a history of the emergence of professional meteorology, see Harper, *Weather by the Numbers*.

11. Liboiron continues, "Settlers do not have to set foot on the Land, own the Land, or even use the Land as Resource so long as the Land is available for settler futures" (*Pollution Is Colonialism*, 66). While Liboiron is discussing access to "Land as Resource" with specific regard to hazardous waste, their point helps to clarify the ways that settler weather data collection, too, assumes unbridled access to land, while also provoking a future where the data collected is processed and stored. Whether on land or in the "cloud," data processing and storage are never materially free. Much as with hazardous waste, data costs the biosphere and its inhabitants at differential levels. On the environmental costs of data centers, see Hogan, "Data Flows and Water Woes"; Hogan, "Facebook Data Storage Centers"; Pasek, "Managing Carbon and Data Flows."

12. On settler futurity, colonialism, and settler colonialism, see Tuck and Yang, "Decolonialism Is Not a Metaphor"; Liboiron, *Pollution Is Colonialism* (esp. pp. 16–17, 36, 65); see also Mitchell and Chaudhury, "Worlding beyond 'the' 'End' of 'the World'"; Rifkin, *Beyond Settler Time*; Strakosch and Macoun, "Vanishing Endpoint of Settler Colonialism."

13. Harper, *Weather by the Numbers*, 12.

14. US Department of Agriculture, Weather Bureau, *Report of the Chief* (1910), 41–55.

15. US Department of Agriculture, Weather Bureau, *Report of the Chief* (1910), 41.

16. Harper, *Weather by the Numbers*, 61.

17. In *Calculating the Weather*, Frederik Nebeker skillfully outlines the ways in which numerical meteorology emerged in the interwar period. Numerical meteorology grounded itself in specific types of data, namely, algorithmic data. Numerical and computer forecasts gained popularity and unified the field of meteorology from the 1950s on, allowing for individual models to be incorporated into larger-scale models, such as climate models.

18. Appadurai, *Modernity at Large*, 116.

19. "Disasters of this kind," Dust Bowl historian Donald Worster argues, "challenge a society's capacity to think—require it to analyze and explain and learn from misfortune" (Worster, *Dust Bowl*, 25).

20. Aronova, "Geophysical Datascapes of the Cold War," 308.

21. Seitz, *Cosmic Inventor*, 17.

22. US Department of Agriculture, Weather Bureau, *Report of the Chief* (1901), 6.

23. Hughes, *Century of Weather Service*, 31.

24. United States and Evans, *Wireless Telegraphy*, 10.

25. US Department of Agriculture, Weather Bureau, *Report of the Chief* (1904), 26.

26. US Department of Agriculture, Weather Bureau, *Report of the Chief* (1907), 34.

27. US Department of Agriculture, Weather Bureau, *Report of the Chief* (1915), 1.

28. US Department of Agriculture, Weather Bureau, *Report of the Chief* (1916), 7.

29. US Department of Agriculture, Weather Bureau, *Report of the Chief* (1916), 7.

30. *Report of the Secretary of Agriculture*, 24.

31. Espy, "Circular in Relation to Meteorological Observations."

32. The "system of merit and discipline" was a civil service law that governed advancement and dismissal more generally at the civil service entities; it was the "American system, the unaristocratic system" meant to replace systems of patronage. See *Report of the United States Civil-Service Commission* (Washington, DC: Government Printing Office, 1887).

33. US Department of Agriculture, Weather Bureau, *Report of the Chief* (1900), 15.

34. Whitnah, *History of the United States Weather Bureau*, 111.

35. Nebeker, *Calculating the Weather*, 87.

36. *United States Congressional Serial Set*.

37. *United States Congressional Serial Set*, 279.

38. *United States Congressional Serial Set*, 280.

39. United States and US Civil Service Commission, *Official Register of the United States*, 1066–70.

40. United States and US Civil Service Commission, *Official Register of the United States*, 1072.

41. *Second Convention of the Weather Bureau Officials.* Under "The Personnel of the Station Force": "An observer of the opposite sex in charge of a regular station of the Bureau (the widow of a veteran 'observer sergeant') being omitted in the count" (36).

42. "Weather Bureau Men as Educators"; "Weather Bureau Men as University Students."

43. "U.S. Civil-Service Examination for Assistant Observer."

44. The bureau ran a cooperative observer network totaling more than forty-five hundred volunteers. While cooperative data was utilized in ways different from principal station data, these numbers suggest that the data transfer landscape was indeed dense—namely, cooperative observer data was used to form climatological records rather than daily forecasts.

45. Alter, "Cooperative Weather Bureau Observers of Utah," 274.

46. As Kim Tolley discusses, in the early twentieth century, girls expressed an interest in studying classics and Latin over science education. This preference was largely due to the fact that women needed Latin to gain entrance into the prestigious colleges that their male counterparts had been attending for some time (Tolley, *Science Education of American Girls*, 157).

47. See Kohlstedt, "Nature, Not Books," 324–52.

48. US Department of Agriculture, *Weather Bureau Topics and Personnel*, August 1915, 6. The association between clean data and observer practice had been with weather data collection since at least 1873, when a Smithsonian official made the following note in a meteorological observer index book in 1873 that George Harper's "registers are all very much soiled" and that his "register for March 1868 is offensively dirty."

49. US Department of Agriculture, *Weather Bureau Topics and Personnel*, August 1915, 3–4.

50. US Department of Agriculture, *Weather Bureau Topics and Personnel*, August 1915, 3–4.

51. US Department of Agriculture, *Weather Bureau Topics and Personnel*, August 1915, 3–4.

52. US Department of Agriculture, *Weather Bureau Topics and Personnel*, January 1917, 2–3; US Department of Agriculture, *Weather Bureau Topics and Personnel*, February 1917, 3.

53. US Department of Agriculture, *Weather Bureau Topics and Personnel*, February 1917, 3.

54. Marvin, United States, and Weather Bureau, *Instructions for Obtaining and Transcribing Records* (1894), 53.

55. Marvin, United States, and Weather Bureau, *Instructions for Obtaining and Transcribing Records* (1894), 5.

56. Marvin, United States, and Weather Bureau, *Instructions for Obtaining and Transcribing Records* (1894), 5.

57. Marvin, United States, and Weather Bureau, *Instructions for Obtaining and Transcribing Records* (1894), 5.

58. Marvin, United States, and Weather Bureau, *Instructions for Obtaining and Transcribing Records* (1894), 7.

59. Marvin, United States, and Weather Bureau, *Instructions for Obtaining and Transcribing Records* (1894), 7.

60. Marvin, United States, and Weather Bureau, *Instructions for Obtaining and Transcribing Records* (1894), 7.

61. Marvin, United States, and Weather Bureau, *Instructions for Obtaining and Transcribing Records* (1903), 25–26.

62. Marvin, United States, and Weather Bureau, *Instructions for Obtaining and Transcribing Records* (1903), 31.

63. Charles F. Marvin served as chief of the Weather Bureau up until his retirement in 1934. Prior to his official role at the Weather Bureau, Marvin was a professor of meteorology at the Weather Bureau; prior to the bureau's 1890 founding, he was a junior professor at the Office of the Chief Signal Officer of the US Army. During this time, Marvin wrote on a variety of meteorological topics, including aerology, instrument care, and data transcription.

64. Marvin, *Instructions for Obtaining and Transcribing Records* (1894), 7–8; Marvin, *Instructions for Obtaining and Transcribing Records* (1915), 5–8.

65. Harper, *Weather by the Numbers*, 58.

66. Harper, *Weather by the Numbers*, 13.

67. Robinson, "Composition and Origin of the Dust," 86.

68. Throughout the early nineteenth century, forts such as Leavenworth in Kansas and Gibson in Oklahoma operated as settler expansion outposts, at which army surgeons were required by US Army regulation to collect weather data at hospital and prison sites. See the Climate Data Modernization Project Map, accessed September 15, 2020, https://mrcc.purdue.edu/data_serv/cdmp/cdmp.jsp.

69. The ancestral territory of the Osage Nation spans from present-day Kansas, Oklahoma, Arizona, Montana, and Kentucky into portions of Illinois, Ohio, Wisconsin, West Virginia, Pennsylvania, Louisiana, Texas, Tennessee, and Colorado. The present-day Osage Reservation is in Oklahoma. To learn more, see the Osage Nation's website, https://www.osageculture.com/culture/geography. The territory of the Kaw included what is present-day northeastern and eastern Kansas. With the Indian Removal Act of 1830, the Kaw were forced to relocate to Oklahoma when "the federal government forcibly transplanted nearly 100,000 people comprising tribes such as the Shawnee, Delaware, Wyandot, Kickapoo, Miami, Sac and Fox, Ottawa, Peoria, and Potawatomie onto lands claimed by the Kaw and Osage." To learn more, see the Kaw Nation's website, https://www.kawnation.gov/cultural-history-part-2. The Comanche "migrated across the Plains, through Wyoming, Nebraska, Colorado, Kansas, New Mexico, Texas, and Oklahoma" and "ultimately settled here in Southwest Oklahoma." The present-day Comanche Nation is located outside of Lawton, Oklahoma. To learn more, see the Comanche Nation's website, https://comanchenation.com/our-nation/about-us.

70. US Department of Agriculture, Weather Bureau, "Climatological Data," 55.

71. Built in 1824, Fort Gibson was the headquarters of the southwestern frontier. Settlers living at the post were illegally occupying the lands of the Osage and Cherokee tribes, the present-day lands of the Sioux and Osage. See Land Cessions 97 and 492, December 2021, https://usfs.maps.arcgis.com/apps/webappviewer/index.html?id=fe311f69cb1d43558227d73bc34f3a32.

72. US Department of Agriculture, Weather Bureau, *Report of the Chief* (1891), 555.

73. "Our Veteran Cooperative Observers," 313; "Instructions for the Cooperative Observers," 1.

74. Cox, "Change in the Method of Transmitting," 78.

75. Although the Weather Bureau began conducting atmospheric dust measurements in 1922 on the campus of American University, these types of measurements did not come in handy for station agents and volunteer observers as dust storms swept across the country throughout the 1930s.

76. US Department of Agriculture, *Weather Bureau Topics and Personnel*, January 1938, 8.

77. US Department of Agriculture, *Weather Bureau Topics and Personnel*, August 1935, 168; 1936, 231.

78. See Cronon, *Changes in the Land*; Worster, *Dust Bowl*.

79. Harper, *Weather by the Numbers*, chaps. 2 and 3.

80. Whitnah, *History of the United States Weather Bureau*, 132.

81. Markus et al., *Air Weather Service*, 4–5.

82. Whitnah, *History of the United States Weather Bureau*, 202–3.

83. "Report of the Science Advisory Board," 18.

84. "Meteorology of the Transatlantic Flights," 122.

85. Harper, *Weather by the Numbers*, 35.

86. Parkinson, "Dust Storms over the Great Plains."

87. Parkinson, "Dust Storms over the Great Plains," 127.

88. Clap, "Brief History," 2.

89. Clap, "Brief History," 2.

90. Harper, *Weather by the Numbers*, 96.

91. Harper, *Weather by the Numbers*, 187–223.

92. Whyte, "Dakota Access Pipeline," 330.

93. Goddard Space Flight Center, *Final Report on the TIROS I*, i.

94. See George H. Kimble, "Tomorrow's Forecasts—Less Cloudy: Predicting the Weather May Be Revolutionized by the Satellites Foreshadowed by TIROS I. Forecasts," *New York Times*, May 1, 1960, SM49.

CHAPTER 5. **Ugly Data in the Age of Satellites and Extreme Weather**
Grateful acknowledgment is made to *American Literature*, where portions of this chapter first appeared. Sara J. Grossman, "Ugly Data in the Age of Weather Satellites," 815–37.

1. "Luncheon Address," 113.

2. "Luncheon Address," 111–12.

3. Whitnah, *History of the United States Weather Bureau*, 33.

4. Whitnah, *History of the United States Weather Bureau*, 170.

5. Fleming, "Telegraphing the Weather," 155.

6. RAND Corporation, Davies, and Harris, *RAND's Role*, 6–8.

7. Douglass Aircraft Company, *Preliminary Design*, 2.

8. James E. Lipp et al., "Reference Papers Relating to a Satellite Study, Project RAND, Douglass Aircraft Company," February 1, 1947, 39.

9. Lipp et al., "Reference Papers Relating to a Satellite Study," 45.

10. RAND Corporation, Greenfield, and Kellogg, "Inquiry into the Feasibility of Weather Reconnaissance," v.

11. Lipp and Salter, "Project Feedback."

12. Chapman, "Case Study of the U.S. Weather Satellite Program," 24.

13. Lipp and Salter, "Project Feedback," 11.

14. Lipp and Salter, "Project Feedback," 5.

15. Lipp and Salter, "Project Feedback," 85.

16. RAND Corporation, Davies, and Harris, *RAND's Role*, 25.

17. RAND Corporation, Davies, and Harris, *RAND's Role*, 25.

18. "Luncheon Address," 114.

19. Goddard Space Flight Center, *Final Report on the TIROS I*, 149; Maher, *Apollo in the Age of Aquarius*, 61.

20. Collins, "Memo." See also Maher, *Apollo in the Age of Aquarius*, 64–65.

21. Butler and Sternberg, "TIROS," 252.

22. Perry, *History of Satellite Reconnaissance*, 218–19.

23. Hall, *History of the Military Polar Orbiting Meteorological Satellite Program*, 11.

24. Central Intelligence Agency, "Memorandum for the Deputy Director: Intelligence Aspects of TIROS," 3.

25. On TIROS as a "shield" for DMSP, see Brugioni, *Eyes in the Sky*, 199; on TIROS as a "peaceful" use of space, see Kalic, *US Presidents and the Militarization of Space*, 51.

26. Prior to TIROS's launch, the CIA raised concern that NASA's intention to release earth images to the public "would provide the Soviets with free information on the effectiveness of the system for reconnaissance" and "may not be desirable." Central Intelligence Agency, "Memorandum for the Deputy Director: Intelligence Aspects of TIROS: Project TIROS," 2.

27. Halpern, *Beautiful Data*, 99.

28. Halpern, *Beautiful Data*, 94. We should also keep in mind that a visualization, after all, does not *re*-present something that has already been seen; it makes visual what "is visualizable but not necessarily visible" (Halpern, *Beautiful Data*, 38).

29. Arnold, *Spying from Space*, 7, 25.

30. Maini and Agrawal, *Satellite Technology*, 371.

31. Parks, *Cultures in Orbit*, 83–84.

32. Goddard Space Flight Center, *Final Report on the TIROS I*. TIROS observed

earth through Hollywood motion picture technology, with "motion picture cameras" mounted to the satellite.

33. George Kimble, "Tomorrow's Forecasts—Less Cloudy: Predicting the Weather May Be Revolutionized by the Satellites Foreshadowed by TIROS I. Forecasts," *New York Times*, May 1, 1960, SM49.

34. Winston, "Satellite Pictures."

35. Winston, "Satellite Pictures," 312.

36. It is worthwhile to consider that in 1961, the US Department of Commerce, under the guidance of Weather Bureau chief F. W. Reichelderfer, published some of TIROS's inaugural findings in a collection titled *Catalogue of Meteorological Satellite Data—TIROS I Television Cloud Photography*. The collection is an index of coverage, a self-titled "catalogue" of "data" belonging to the satellite pictures themselves. Data included orbit number, frame, reel, and start time. This document is interesting because it offers a record of the picture data itself, archiving every single televisual move that the satellite made. The document is a chronology of pictorial data—a kind of early metadata that underscores the complexity of TIROS pictures as pure outputs of vision.

37. TIROS 8 was launched on December 21, 1963, with the first APT TV direct readout. Schnapf and McElroy, *Monitoring Earth's Ocean*, 53–59.

38. Schnapf and McElroy, *Monitoring Earth's Ocean*, 78.

39. Mack, *Viewing the Earth*, 102.

40. Schnapf and McElroy, *Monitoring Earth's Ocean*, 74.

41. Williamson, *Spacecraft Technology*, 149.

42. Mack, *Viewing the Earth*, 68; Kramer, *Observation of the Earth*, 721.

43. Schnapf and McElroy, *Monitoring Earth's Ocean*, 77.

44. Mack, *Viewing the Earth*, 70–71. MSS was developed by the Infrared Laboratory at the University of Michigan, a lab founded in 1946 with a contract from the Department of Defense to investigate ballistic missiles and, in 1954–55, reconnaissance and surveillance of battlefields (Mack, *Viewing the Earth*, 71).

45. Mack, *Viewing the Earth*, 81–93.

46. Maher, *Apollo in the Age of Aquarius*, 77.

47. Mack, *Viewing the Earth*, 174.

48. Mack, *Viewing the Earth*, 172.

49. Mack, *Viewing the Earth*, 175.

50. Brant, "Report," 887–89.

51. Brant, "Report," 887–89.

52. Abrams et al., *Joint NASA/Geostat Test Case Project: Final Report*, iii.

53. Abrams et al., *Joint NASA/Geostat Test Case Project: Final Report*, iii.

54. Brant, "Report," 887.

55. Abrams et al., *Joint NASA/Geostat Project: Executive Summary*, 3.

56. Abrams et al., *Joint NASA/Geostat Project: Executive Summary*, 19.

57. Abrams et al., *Joint NASA/Geostat Project: Executive Summary*, 19. The Shoshone-Bannock Tribes, which now live at the Fort Hall Reservation in southeastern Idaho, "occupied vast regions of land encompassing present-day Idaho,

Oregon, Nevada, Utah, Wyoming, Montana, and into Canada." The Fort Hall Reservation was established through the 1868 Fort Bridger Treaty and subsequently was reduced in size in 1872. For more information, see the Shoshone-Bannock Tribes website, http://www.sbtribes.com/about. The Eastern Shoshone, who now reside on the Wind River Reservation in Wyoming, have been living "in the Wind River mountain range and its environs for some 12,000 years." Through the 1863 and 1868 treaties of Fort Bridger, the boundaries of the Eastern Shoshone's lands were redrawn by the US settler states, reducing their acreage to 2,774,400 acres in west central Wyoming. For more information, see the Eastern Shoshone Tribe website, https://easternshoshone.org/about.

58. Abrams et al., *Joint NASA/Geostat Project: Executive Summary*, 26.

59. Cirac-Claveras, "Weather Satellites," 96–97.

60. Cirac-Claveras, "Weather Satellites," 96–97.

61. This included ESSA, ITOS, and NOAA satellites. Schnapf and McElroy, *Monitoring Earth's Ocean*, 53.

62. National Center for Atmospheric Research (US), Advanced Study Program, et al., *Weather Forecasting and Weather Forecasts*, 599.

63. National Advisory Committee on Oceans and Atmosphere, *Report to: The President and the Congress*, 111–13; Harris, *Opportunities in Meteorology*, 136.

64. Harris, *Opportunities in Meteorology*, 136.

65. US Army Corps of Engineers, "Environmental Assessment, Dakota Access Pipeline Project," 728.

66. United States and Genevieve Walker, *Final Supplemental Environmental Impact Statement*, sec. 2.4.2.

67. *Rosebud Sioux Tribe and Fort Belknap Indian Community v. United States Department of State*.

68. Minnesota Department of Commerce Energy Environmental Review and Analysis, "Draft Environmental Impact Statement," chaps. 5 and 7.

69. Henson, *Weather on the Air*, 4.

70. Henson, *Weather on the Air*, 92.

71. Henson, *Weather on the Air*, 93.

72. Henson, *Weather on the Air*, 93

73. Henson, *Weather on the Air*, 94.

74. Henson, *Weather on the Air*, 94.

75. The movie is a compilation of NOAA images and NASA animation.

76. The same was true of Hurricane Katrina, though the nature of mediation was markedly different. As Christina Sharpe notes, images of Black suffering during Hurricane Katrina "work[ed] to solidify and make continuous the colonial project of violence." They "function[ed] as a hail to the non Black person . . . to confirm the status, location, and already held opinions within dominant ideology about those exhibitions of spectacular Black bodies whose meanings then remain unchanged" (Sharpe, *In the Wake*, 116). See also Dyson, *Come Hell or High Water*; Kellner, "Katrina Hurricane Spectacle," 222–34.

77. Wark, "Weird Global Media Event," 266.

78. Monmouth University, "Interview with Patricia Armstrong."

79. Launched in 1960, TIROS was the world's first weather satellite.

80. Davis, "History of NOAA Satellite Programs," 12.

81. Schmit et al., "Geostationary Operational Environmental Satellite," 12.

82. Schmit et al., "Geostationary Operational Environmental Satellite," 6.

83. Schmit et al., "Geostationary Operational Environmental Satellite," 12.

84. Sargent, "NASA's Time-Lapse Video."

85. On settler time, see Rifkin, *Beyond Settler Time*.

86. Guernsey, "The Infrastructures of White Settler Perception," 599; Whyte, "Indigenous Climate Change Studies," 152–58.

87. Parks, *Cultures in Orbit*, 3.

88. Hui Kyong Chun, "Crisis, Crisis, Crisis," 107.

89. Kimmerer, *Braiding Sweetgrass*, 346.

90. Viégas and Wattenberg, *Wind Map*.

91. Oremus, "Web's Best Visualizations of Hurricane Sandy"; Jardin, "Wind Map Shows Sandy's Gusts."

92. Rosenzweig et al., "Developing Coastal Adaptation to Climate Change," 93–127.

93. Barthes, *Camera Lucida*, 87, 89.

94. Sontag, *On Photography*, 10.

95. Butler, *Frames of War*, 67.

96. Drucker, *Graphesis*, 125.

97. Drucker, *Graphesis*, 133.

98. Tufte, *Visual Display*, 154.

99. Tufte, *Visual Display*, 177.

100. Davis and Todd, "On the Importance of a Date."

101. Buell, *Environmental Imagination*.

102. Monmouth University, interview with Patricia Armstrong, in "Oral Histories about Super Storm Sandy."

103. Monmouth University, "Oral Histories about Super Storm Sandy."

104. Buell, *Environmental Imagination*, 2.

105. The storm event was a "Frankenstorm," brought into being by both "natural and unnatural" forces, as Bill McKibbon and scientist Charles H. Green agreed. The increased levels of greenhouse gases have fundamentally altered Arctic sea ice, leading to an increase in extreme weather events like Hurricane Sandy. See also Trenberth, Fasullo, and Shepherd, "Attribution of Climate Extreme Events," 725–30.

106. Malm and Hornborg, "Geology of Mankind?," 62–69.

107. Gross, *Anishinaabe Ways of Knowing and Being*, 33; Whyte, "Indigenous Climate Change Studies," 320.

108. Garrison, "Settler Responsibility," 57.

109. "Tools for Addressing Chapter Conflict," 5.

110. Tuck and Yang, "Decolonization Is Not a Metaphor," 35.

111. brown, *Emergent Strategy*, 79.

112. Bell et al., "Enacting Settler Responsibilities towards Decolonisation," 8.

113. "Tools for Addressing Chapter Conflict," 8.

114. Tuck and Yang, "Decolonization Is Not a Metaphor," 36.

EPILOGUE

1. Maphanyane, "Parallel Development of Three Major Space Technology Systems and Human Side of Information Reference Services as an Essential Complementary Method," 3491.

2. The GOES 1 satellite—launched in 1974—was decommissioned in 1985; GOES 2—launched in 1977—was decommissioned in 1993 and reactivated in 1995 only to be decommissioned again in 2001. And GOES 3—launched in 1978—was decommissioned in 2016, having operated for twenty-eight years as a weather and a communications satellite.

3. Barnes et al., *TIROS-N Series Direct Readout Services Users Guide*.

4. Leese et al., *Archiving and Climatological Applications*, D-2.

5. National Research Council (US) and Space Science Board, *Data Management and Computation*, 43.

6. Climate Change Science Program (US) and Subcommittee on Global Change Research, *Strategic Plan*.

7. Climate Change Science Program (US) and Subcommittee on Global Change Research, *Strategic Plan*, 9.

8. Cass, "Grief Clause," 38.

9. See Harjo, "Map to the Next World."

10. Leiserowitz et al., *Climate Change in the American Mind*.

11. Arvin, Tuck, and Morrill, "Decolonizing Feminism," 9.

12. Olhoff and Christensen, *Emissions Gap Report 2019*, xx.

TRIBAL, NATION, AND CONFEDERACY WEBSITES

Association of Ramaytush Ohlone. Accessed November 12, 2020. https://www
.ramaytush.org/ramaytush-ohlone.html.
Cayuga Nation. "People of the Great Swamp." Accessed June 26, 2022. http:/
/cayuganation-nsn.gov.
Comanche Nation. Accessed July 3, 2022. https://comanchenation.com.
Eastern Shoshone Tribe. Accessed February 17, 2023. https://easternshoshone
.org/about.
Haudenosaunee Confederacy. "Land Rights and Treaties." Accessed June 26, 2022.
https://www.haudenosauneeconfederacy.com.
Hopi Tribe. Accessed June 29, 2022. https://www.hopi-nsn.gov.
Kaw Nation. Accessed July 3, 2022. https://kawnation.com.
Lenape Nation of Pennsylvania. Accessed July 15, 2022. https://www.lenape
-nation.org.
Omaha Tribal Historical Research Project. Accessed June 27, 2022. http://
jackalopearts.org/othrp.htm.
Osage Nation. Accessed November 25, 2020. https://www.osageculture.com.
Seminole Tribe of Florida. Accessed June 29, 2022. https://www.semtribe.com
/stof.
Shoshone-Bannock Tribes. Accessed February 17 2023. http://www.sbtribes.com
/about.
White Mountain Apache Tribe. Accessed June 29, 2022. http://www.wmat.us.

BOOKS, JOURNALS, AND ADDITIONAL SOURCES

Abbate, Janet. *Recoding Gender: Women's Changing Participation in Computing.*
Cambridge, MA: MIT Press, 2012.
Abrams, M. J., et al. *The Joint NASA/Geostat Test Case Project: Executive Summary.*
Tulsa, OK: American Association of Petroleum Geologists, 1985.
Abrams, M. J., et al. *The Joint NASA/Geostat Test Case Project: Final Report.* Pt.
2, vols. 1 and 2. Pt. 3, Plates. Tulsa, OK: American Association of Petroleum
Geologists, 1985.
"Aerial Navigation." *Scientific American* 69, no. 8 (August 19, 1893): 309–10.

Ahmed, Sara. "A Phenomenology of Whiteness." *Feminist Theory* 8, no. 2 (August 2007): 149–68.

Alter, Cecil. "The Cooperative Weather Bureau Observers of Utah." *Monthly Weather Review* 40, no. 2 (February 1912): 272–74.

American Association for the Advancement of Science. *Proceedings from the Meeting.* Philadelphia, 1848.

Anderson, Benedict. *Imagined Communities: Reflections on the Origin and Spread of Nationalism.* London: Verso, 2006.

Anderson, Katharine. *Predicting the Weather: Victorians and the Science of Meteorology.* Chicago: University of Chicago Press, 2005.

Appadurai, Arjun. *Modernity at Large: Cultural Dimensions of Globalization.* Minneapolis: University of Minnesota Press, 2003.

Arnold, David C. *Spying from Space: Constructing America's Satellite Command and Control Systems.* College Station: Texas A&M University Press, 1996.

Aronova, Elena. "Geophysical Datascapes of the Cold War: Politics and Practices of the World Data Centers in the 1950s and 1960s." *Osiris* 32, no. 1 (September 2017): 307–27.

"Artificial Production of Rain, The." *Scientific American* 65, no. 10 (September 5, 1891): 145.

Arvin, Maile, Eve Tuck, and Angie Morrill. "Decolonizing Feminism: Challenging Connections between Settler Colonialism and Heteropatriarchy." *Feminist Formations* 25, no. 1 (2013): 8–34.

Barker, Joanne. "Territory as Analytic: The Dispossession of Lenapehoking and the Subprime Crisis." *Social Text* 36, no. 2 (135) (June 2018): 19–39.

Barnes, James C., Michael D. Smallwood, United States, National Earth Satellite Service, and Inc. Environmental Research and Technology. *TIROS-N Series Direct Readout Services Users Guide.* Washington, DC: The Service, 1982.

Barthes, Roland. *Camera Lucida: Reflections on Photography.* New York: Hill and Wang, 1981.

Baumgartner, Kabria. *In Pursuit of Knowledge: Black Women and Educational Activism in Antebellum America.* New York: New York University Press, 2019.

Beattie, James, Emily O'Gorman, and Matthew Henry, eds. *Climate, Science, and Colonization Histories from Australia and New Zealand.* New York: Palgrave Macmillan US, 2016.

Beck, Theodric Romeyn, and Joseph Henry. *An Abstract of the Returns of Meteorological Observations Made to the Regents of the University, by Sundry Academies in This State, in Obedience to Instructions, Dated March 1, 1825.* Albany, NY: n.p., 1829.

Bell, Avril, Rose Yukich, Billie Lythberg, and Christine Woods. "Enacting Settler Responsibilities towards Decolonization." *Ethnicities* 22, no. 5 (2021): 1–14.

Bishop, William D., Joseph Henry, Franklin B. Hough, and James H. Coffin. *Results of Meteorological Observations, Made under the Direction of the United States Patent Office and the Smithsonian Institution from the Year 1854 to 1859, Inclusive, Being a Report of the Commissioner of Patents Made at the First Session of the Thirty-Sixth Congress*. Washington, DC: Government Printing Office, 1861.

Blodget, Lorin. *Climatology of the United States*. Philadelphia: Lippincott, 1857.

Blum, Hester. *The News at the Ends of the Earth: The Print Culture of Polar Exploration*. Durham, NC: Duke University Press, 2019.

Boylan, Anne. "Claiming Visibility: Women in Public / Public Women in the United States, 1865–1910." In *Becoming Visible: Women's Presence in Late Nineteenth-Century America*, edited by Janet Floyd, R. J. Ellis, and Lindsey Traub, 15–40. Amsterdam: Rodopi, 2010.

Brant, Arthur A. "Report: The Geostat Committee, Inc." *Geophysics* 42, no. 4 (June 1977), 715–897.

Brimmner, Larry D. *The Rain Wizard: The Amazing, Mysterious, True Life of Charles Mallory Hatfield*. Honesdale, PA: Calkins Creek, 2015.

brown, adrienne maree. *Emergent Strategy: Shaping Change, Changing Worlds*. Chico, CA: AK Press, 2017.

brown, adrienne maree. *Grievers*. Chico, CA: AK Press, 2021.

Brugioni, Dino A. *Eyes in the Sky: Eisenhower, the CIA, and Cold War Aerial Espionage*. Annapolis, MD: Naval Institute Press, 2011.

Buell, Lawrence. *The Environmental Imagination: Thoreau, Nature Writing, and the Formation of American Culture*. Cambridge, MA: Belknap Press of Harvard University Press, 1995.

Bullard, Robert D. "The Threat of Environmental Racism." *Natural Resources and Environment* 7, no. 3 (1993): 23–56.

Butler, H. I., and S. Sternberg. "TIROS—The System and Its Evolution." *IRE Transactions on Military Electronics* MIL-4, no. 2/3 (1960): 248–56.

Butler, Judith. *Frames of War: When Is Life Grievable?* London: Verso, 2009.

Cass, Meg. "Grief Clause." *Ecotone* 16, no. 2 (2020): 36–47.

Chapman, Richard LeRoy. "A Case Study of the U.S. Weather Satellite Program: The Interaction of Science and Politics." PhD diss., Syracuse University, 1967.

Central Intelligence Agency. "Memorandum for the Deputy Director: Intelligence Aspects of TIROS: Project TIROS." July 13, 1959.

Central Intelligence Agency. "Memorandum for the Deputy Director: Intelligence Aspects of TIROS." July 17, 1959.

Ceruzzi, Paul E. "When Computers Were Human." *IEEE Annals of the History of Computing* 13, no. 3 (1991): 237–44.

Chudacoff, Howard P. *Children at Play: An American History*. New York: New York University Press, 2007.

Cirac-Claveras, Gemma. "Weather Satellites: Public, Private and Data Sharing. The Case of Radio Occultation Data." *Space Policy* 47 (2019): 94–106.

Clap, Philip F. "A Brief History of the Extended Forecast Division." Meteorological Satellite Laboratory, 1977.

Climate Change Science Program (US) and Subcommittee on Global Change Research. *Strategic Plan for the Climate Change Science Program*. Washington, DC: US Climate Change Science Program, 2003.

Clyde, John Cunningham. *Life of James H. Coffin: For Twenty-Seven Years Professor of Mathematics and Astronomy in Lafayette College; Member of the National Academy of Sciences, and Other Learned Bodies; Discoverer of the Laws Which Govern the Winds of the Globe*. Easton, PA: n.p., 1881.

Coffin, James Henry. *Meteorological Register and Scientific Journal*. Ogdensburg, NY: n.p., 1839.

Coffin, James Henry, Aleksandr Ivanovich Voeïkov, and Selden Jennings Coffin. *The Winds of the Globe: Or, the Laws of Atmospheric Circulation over the Surface of the Earth*. Washington, DC: Smithsonian Institution, 1875.

Collins, Edward M. "Memo, Office of the Assistant Chief of Staff, National Reconnaissance Office: Space Espionage Plans and International Law." March 8, 1961.

Conover, John H. *The Blue Hill Meteorological Observatory: The First 100 Years, 1885–1985*. Boston: American Meteorological Society, 1990.

Cox, Henry J. "Change in the Method of Transmitting Weather Observations." *Monthly Weather Review* 9, no. 4 (April 1928): 77–79.

Cronon, William. *Changes in the Land*. New York: Farrar, Straus and Giroux, 2013.

Cullen, Beth, and Christina Leigh Geros. "Constructing the Monsoon: Colonial Meteorological Cartography, 1844–1944." *History of Meteorology* 9 (2020): 1–26.

"Danger from Street Telegraph Wires." *Scientific American* 34, no. 18 (April 29, 1876): 278.

Daniels, George H. *The Process of Professionalization in American Science: The Emergent Period, 1820–1860*. Cambridge, MA: History of Science Society, 1967.

Daston, Lorraine. "Super-Vision: Weather Watching and Table Reading in the Early Modern Royal Society and Académie Royale Des Sciences." *Huntington Library Quarterly* 78, no. 2 (2015): 187–215.

Daston, Lorraine, and Elizabeth Lunbeck, eds. *Histories of Scientific Observation*. Chicago: University of Chicago Press, 2011.

Davis, Gary. "History of NOAA Satellite Programs," NOAA Satellite and Information Service, October 2009.

Davis, Heather, and Zoe Todd. "On the Importance of a Date, or, Decolonizing the Anthropocene." *ACME: An International Journal for Critical Geographies* 16, no. 4 (December 2017): 761–80.

D'Ignazio, Catherine, and Lauren F. Klein. *Data Feminism*. Cambridge, MA: MIT Press, 2020.

Douglass Aircraft Company. *Preliminary Design of a World-Circulating Space Ship*. Santa Monica, CA: RAND Corporation, 1946.

Drucker, Johanna. *Graphesis: Visual Forms of Knowledge Production*. Cambridge, MA: Harvard University Press, 2014.

Dunaway, Finis. *Seeing Green: The Use and Abuse of American Environmental Images*. Chicago: University of Chicago Press, 2015.

Dunbar-Ortiz, Roxanne. *An Indigenous Peoples' History of the United States*. Boston: Beacon Press, 2015.

Dyson, Michael. *Come Hell or High Water: Hurricane Katrina and the Color of Disaster.* New York: Basic Civitas, 2006.

Eddy, Ruth Story Devereux. *The Eddy Family in America: A Genealogy Comp. by Ruth Story Devereux Eddy . . . and Pub. under the Direction of the Eddy Family Association, in Commemoration of the Three Hundredth Anniversary of the Landing of John and Samuel Eddy at Plymouth, Oct. 29, 1630.* Boston: T. O. Metcalf, Will Currier Eddy Associate, 1930.

Eddy, W. A. "A Record of Some Kite Experiments." *Monthly Weather Review* 26, no. 10 (October 1898): 450–52.

Eddy, W. A. "Science and Kite Flying." *Independent* 52 (September 1900): 2333.

Eddy, W. A., and H. H. Clayton. "Eddy Malay Tailless Kite," *Scientific American* 71, no. 11 (September 15, 1894).

Edwards, Paul. *A Vast Machine: Computer Models, Climate Data, and the Politics of Global Warming.* Cambridge, MA: MIT Press, 2010.

Environmental Document Access and Display System, Version 2. "Dutch Log Books." Accessed June 26, 2022. https://www.ncdc.noaa.gov/EdadsV2.

Environmental Document Access and Display System, Version 2. "East India Company Logbooks." Accessed June 26, 2022. https://www.ncdc.noaa.gov/EdadsV2.

Environmental Document Access and Display System, Version 2. "Iraq Upper Air Data." Accessed June 26, 2022. https://www.ncdc.noaa.gov/EdadsV2.

Epler, Percy H. *The Life of Clara Barton.* New York: Macmillan, 1915.

Espy, James P. "Circular to the Friends of Science." In *United States Congressional Serial Set*, Washington, DC: US Congress, 1843.

Espy, James P. *First Report on Meteorology, to the Surgeon General of the United States Army.* Washington City [sic]: US Army, 1843.

Espy, James P. "On the Importance of Hygrometric Observations in Meteorology, and the Means of Making Them with Accuracy." *Journal of the Franklin Institute* 11, no. 4 (April 1, 1831): 221–29.

Espy, James P. *The Philosophy of Storms, by James P. Espy.* London: H. G. Bohn, 1841.

Espy, James P. *Second Report on Meteorology, to the Secretary of the Navy.* Washington, DC: US Army, 1851.

Espy, James P. "To the Honorable." *Daily National Intelligencer.* Washington, DC: Gales and Seaton, 1837.

Fergusson, S. P. "Exploration of the Air by Means of Kites." *Annals of the Astronomical Observatory at Harvard College* 42, no. 1 (1897): 41–75.

Fergusson, S. P. "Kite Experiments at the Blue Hill Meteorological Observatory." *Monthly Weather Review* 24, no. 9 (September 1896): 232–367.

Finney, Carolyn. "Birth of a Nation." In *Black Faces, White Spaces: Reimagining the Relationship of African Americans to the Great Outdoors.* Chapel Hill: University of North Carolina Press, 2014.

Fixico, Donald L. *Bureau of Indian Affairs.* Santa Barbara, CA: ABC-CLIO, 2012.

Fleming, James R. *Meteorology in America, 1800–1870.* Baltimore: Johns Hopkins University Press, 1990.

Fleming, James R. "Telegraphing the Weather: Military Meteorology, Strategy, and 'Homeland Security' on the American Frontier in the 1870s." In *Instrumental in War: Science, Research, and Instruments between Knowledge and the World*, edited by Steven Walton, 153–78. Boston: Brill, 2005.

"Flying Machines in America." *Aeronautical Journal* 3, no. 9 (1899): 13–14.

Gabrys, Jennifer. *Program Earth: Environmental Sensing Technology and the Making of a Computational Planet*. Minneapolis: University of Minnesota Press, 2017.

Garrison, Rebekah. "Settler Responsibility: Respatialising Dissent in 'America' beyond Continental Borders." *Shima Journal* 13, no. 2 (2019): 56–75.

"Gigantic Kite, A." *Scientific American* 55, no. 20 (November 13, 1886): 310.

Gitelman, Lisa. *Always Already New: Media, History, and the Data of Culture*. Cambridge, MA: MIT Press, 2006.

Gitelman, Lisa. *Paper Knowledge: Toward a Media History of Documents*. Durham, NC: Duke University Press, 2014.

Gitelman, Lisa, ed. *"Raw Data" Is an Oxymoron*. Cambridge, MA: MIT Press, 2013.

Goddard Space Flight Center. *Final Report on the TIROS I Meteorological Satellite System*. Washington, DC: National Aeronautics and Space Administration, 1962.

Golinski, Jan. *British Weather and the Climate of Enlightenment*. Chicago: University of Chicago Press, 2007.

Golinski, Jan. "The Care of the Self and the Masculine Birth of Science." *History of Science* 40, no. 2 (2002): 125–45.

Gotkin, Kevin. "When Computers Were Amateur." *IEEE Annals of the History of Computing* 36, no. 2 (2014): 4–14.

Gross, Lawrence William. *Anishinaabe Ways of Knowing and Being*. London: Routledge, 2016.

Grossman, Sara J. "Gendering Nineteenth-Century Data: The Women of the Smithsonian Meteorological Project." *Journal of Women's History* 33, no. 1 (2021): 85–109.

Grossman, Sara J. "Ugly Data in the Age of Weather Satellites." *American Literature* 88, no. 4 (2016): 815–37.

Guernsey, Paul. "The Infrastructures of White Settler Perception: A Political Phenomenology of Colonialism, Genocide, Ecocide, and Emergency." *Nature and Space* 5, no. 2 (2021): 588–604.

Haffner, Jeanne. *The View from Above: The Science of Social Space*. Cambridge, MA: MIT Press, 2013.

Hall, R. Cargil. *A History of the Military Polar Orbiting Meteorological Satellite Program*. Chantilly, VA: National Reconnaissance Office, 2001.

Halpern, Orit. *Beautiful Data: A History of Vision and Reason Since 1945*. Durham: Duke University Press, 2014.

Hardy, Penelope. "Meteorology as Nationalism on the German Atlantic Expedition." *History of Meteorology* 8 (2017): 124–44.

Harjo, Joy. "A Map to the Next World." In *How We Became Human: New and Selected Poems, 1975–2001*, 129–32. New York: Norton, 2004.

Harper, Kristine. *Weather by the Numbers: The Genesis of Modern Meteorology.* Cambridge, MA: MIT Press, 2008.

Harris, Cheryl I. "Whiteness as Property." *Harvard Law Review* 106, no. 8 (1993): 1707–91.

Harris, Miles. *Opportunities in Meteorology.* New York: Universal Publishing, 1972.

"Harvey-Level Damage Probably Won't Happen in Philadelphia, but Intense Flooding Already Does." State Impact Pennsylvania. September 1, 2017. https://stateimpact.npr.org/pennsylvania/2017/09/01/harvey-level-damage -probably-wont-happen-in-philadelphia-but-intense-flooding-already-does/.

Hauptman, Laurence M. *Conspiracy of Interests: Iroquois Dispossession and the Rise of New York State.* Syracuse, NY: Syracuse University Press, 2001.

Hauptman, Laurence M., and L. Gordon McLester. *The Oneida Indian Journey from New York to Wisconsin, 1784–1860.* Madison: University of Wisconsin Press, 1999.

Heglar, Mary Annaïse. "Climate Change Isn't the First Existential Threat." ZORA (blog), February 18, 2020. https://zora.medium.com/sorry-yall-but-climate -change-ain-t-the-first-existential-threat-b3c999267aa0.

Henry, Joseph. "Programme of Organization." In *Annual Report of the Board of Regents of the Smithsonian Institution.* Washington, DC: Government Printing Office, 1847.

Henry, Joseph. *Scientific Writings of Joseph Henry.* Washington, DC: Smithsonian Institution, 1886.

Henry, Joseph, and James Espy. "Circular from the Smithsonian Institution, Relative to Meteorological Observations." Smithsonian Institution Archives, Record Unit 60, Smithsonian Institution, Meteorological Project, Records.

Henry, Joseph. *The Papers of Joseph Henry.* Vol. 1, *December 1797–October 1832: The Albany Years,* edited by Nathan Reingold, Stuart Pierson, and Arthur P. Molella with the assistance of James M. Hobbins and John R. Kerwood. Washington, DC: Smithsonian Institution Press, 1972.

Henry, Joseph. *The Papers of Joseph Henry.* Vol. 2, *The Princeton Years: November 1832–December 1835,* edited by Nathan Reingold, Arthur P. Molella, and Michele L. Aldrich with the assistance of James M. Hobbins and Kathleen Waldenfels. Washington, DC: Smithsonian Institution Press, 1975.

Henry, Joseph. *The Papers of Joseph Henry.* Vol. 3, *January 1836–December 1837: The Princeton Years,* edited by Nathan Reingold, Arthur P. Mollella, and Marc Rothenburg with the assistance of Kathleen Waldenfels and Joan N. Bodansky. Washington, DC: Smithsonian Institution Press, 1979.

Henry, Joseph. *The Papers of Joseph Henry.* Vol. 4, *January 1838–December 1840: The Princeton Years,* edited by Nathan Reingold, Arthur P. Mollella, and Marc Rothenburg with the assistance of Kathleen Waldenfels and Joan F. Steiner. Washington, DC: Smithsonian Institution Press, 1981.

Henry, Joseph. *The Papers of Joseph Henry.* Vol. 5, *January 1841–December 1843: The Princeton Years,* edited by Marc Rothenburg and Kathleen W. Dorman. Washington, DC: Smithsonian Institution Press, 1985.

Henry, Joseph. *The Papers of Joseph Henry.* Vol. 6, *January 1844–December 1846: The Princeton Years*, edited by Marc Rothenburg, Kathleen W. Dorman, John C. Rumm, and Paul H. Theerman. Washington, DC: Smithsonian Institution Press, 1992.

Henry, Joseph. *The Papers of Joseph Henry.* Vol. 7, *January 1847–December 1849: The Smithsonian Years*, edited by Marc Rothenburg, Paul H. Theerman, Kathleen W. Dorman, John C. Rumm, and Deborah Y. Jeffries. Washington, DC: Smithsonian Institution Press, 1996.

Henry, Joseph. *The Papers of Joseph Henry.* Vol. 8, *January 1850–December 1853: The Smithsonian Years*, edited by Marc Rothenburg, Kathleen W. Dorman, Frank R. Millikan, and Deborah Y. Jeffries. Washington, DC: Smithsonian Institution Press, 1998.

Henry, Joseph. *The Papers of Joseph Henry.* Vol. 9, *January 1854–December 1857: The Smithsonian Years*, edited by Marc Rothenburg, Kathleen W. Dorman, Frank R. Millikan, and Deborah Y. Jeffries. Sagamore Beach, MA: Smithsonian Institution in association with Science History Publications, 2002.

Henry, Joseph. *The Papers of Joseph Henry.* Vol. 10, *January 1858–December 1865: The Smithsonian Years*, edited by Marc Rothenburg, Kathleen W. Dorman, Frank R. Millikan, Deborah Y. Jeffries, and Sarah Shoenfeld. Sagamore Beach, MA: Smithsonian Institution in association with Science History Publications, 2004.

Henson, Robert. *Weather on the Air: A History of Broadcast Meteorology.* Boston: American Meteorological Society, 2010.

Heymann, Matthias. "The Evolution of Climate Ideas and Knowledge." *WIRES Climate Change* 1, no. 4 (2010): 581–97.

Hicks, Mar. *Programmed Inequality: How Britain Discarded Women Technologists and Lost Its Edge in Computing.* Cambridge, MA: MIT Press, 2017.

Hitt, Kristin. "Mauna Kea Protests Reveal Problems in Construction Processes on Native Lands." *Yes Magazine* (August 14, 2019).

Hoffert, Sylvia D. "Female Self-Making in Mid-Nineteenth-Century America." *Journal of Women's History* 20, no. 3 (2008): 34–59.

Hogan, Mél. "Data Flows and Water Woes: The Utah Data Center." *Big Data and Society* 2, no. 2 (2015): 1–12.

Hogan, Mél. "Facebook Data Storage Centers as the Archive's Underbelly." *Television and New Media* 16, no. 1 (January 2015): 3–18.

Howe, Joshua. *Behind the Curve.* Seattle: University of Washington Press, 2016.

Hughes, Patrick. *A Century of Weather Service: A History of the Birth and Growth of the National Weather Service, 1870–1970.* New York: Gordon and Breach, 1970.

Hui Kyong Chun, Wendy. "Crisis, Crisis, Crisis, or Sovereignty and Networks." *Theory, Culture and Society* 28, no. 6 (2011): 91–112.

"Instructions for Meteorological Reports, 1825." In *Instructions from the Regents of the University, to the Several Academies Subject to Thier [sic] Visitation.* Albany, NY: Van Benthuysen and Burt, 1845.

"Instructions for the Cooperative Observers of the Weather Bureau." Washington, DC: Government Printing Office, 1924.

James H. Coffin Papers, 1829–1911. Special Collections and College Archives, Skillman Library, Lafayette College, Easton, PA.

Jankovic, Vladimir. *Reading the Skies: A Cultural History of English Weather, 1650–1820*. Chicago: University of Chicago Press, 2001.

"Japanese Bird Kite, The." *Scientific American* 27, no. 17 (October 26, 1872): 258.

Jardin, Xeni. "Wind Map Shows Sandy's Gusts." *BoingBoing*, October 29, 2012.

Jones, Thomas, and Franklin Institute. *Journal of the Franklin Institute*. Philadelphia: Printed by the Franklin Institute, at their Hall, vol. 7 (1831).

Jones, Thomas, and Franklin Institute. *Journal of the Franklin Institute*. Philadelphia: Printed by the Franklin Institute, at their Hall, vol. 18 (1834).

Jones, Thomas, and Franklin Institute. *Journal of the Franklin Institute*. Philadelphia: Printed by the Franklin Institute, at their Hall, vol. 19 (1837).

"Journal of the Proceedings of the Regents of the Smithsonian Institution." In US Congress, *Miscellaneous Documents: 30th Congress, 1st Session–49th Congress, 1st Session*. Washington, DC: Tippin and Streeper, 1849: 6–54.

Jurin, Jacobo. "Invitatio ad observationes meteorologicas communi consilio instituendas." M. D. Soc. Reg. Secr. & Colleg. Med. Lond. Socio. *Philosophical Transactions (1683–1775)* 32 (1722): 422–27.

Kalic, S. N. *US Presidents and the Militarization of Space, 1946–1967*. College Station: Texas A&M University Press, 2012.

Kelley, Mary. *Learning to Stand and Speak: Women, Education, and Public Life in America's Republic*. Chapel Hill: University of North Carolina Press, 2008.

Kellner, Douglass. "The Katrina Hurricane Spectacle and Crisis of the Bush Presidency." *Cultural Studies ↔ Critical Methodologies* 7, no. 2 (2007): 222–34.

Kidwell, Peggy Aldrich. "American Scientists and Calculating Machines—From Novelty to Commonplace." IEEE *Annals of the History of Computing* 12, no. 1 (1990): 31–40.

Kimmerer, Robin Wall. *Braiding Sweetgrass*. Minneapolis: Milkweed Editions, 2016.

Kimmerer, Robin Wall. "Weaving Traditional Ecological Knowledge into Biological Education: A Call to Action." *BioScience* 52, no. 5 (2002): 432–38.

King, Tiffany Lethabo. *The Black Shoals: Offshore Formations of Black and Native Studies*. Durham, NC: Duke University Press, 2019.

Kitchin, Rob. *The Data Revolution: Big Data, Open Data, Data Infrastructures and Their Consequences*. Los Angeles: Sage, 2014.

Kohlstedt, Sally Gregory. "In from the Periphery: American Women in Science, 1830–1880." *Signs: Journal of Women in Culture and Society* 4, no. 1 (1978): 81–96.

Kohlstedt, Sally Gregory. "Nature, Not Books: Scientists and the Origins of the Nature-Study Movement in the 1890s." *Isis* 96, no. 3 (September 2005): 324–52.

Kouri, Scott, and Hans Skott-Myhre. "Catastrophe: A Transversal Mapping of

Colonialism and Settler Subjectivity." *Settler Colonial Studies* 6, no. 3 (July 2, 2016): 279–94.

Kramer, H. J. *Observation of the Earth and Its Environment: Survey of Missions and Sensors*. Berlin: Springer, 2002.

Krause, Monika. "Reporting and the Transformations of the Journalistic Field: US News Media, 1890–2000." *Media, Culture and Society* 33, no. 1 (2011): 89–104.

Langley, Samuel. "Experiments in Aerodynamics." Washington, DC: Smithsonian Institution, 1891.

Lee, Shetterly M. *Hidden Figures: The Untold Story of the African American Women Who Helped Win the Space Race*. London: William Collins, 2016.

Leese, John A., Arthur L. Booth, Fredric A. Godshall, National Environmental Satellite Center (US), World Meteorological Organization, and Commission for Climatology. *Archiving and Climatological Applications of Meteorological Satellite Data*. Washington, DC: National Environmental Satellite Center, 1970.

Leiserowitz, A., E. Maibach, S. Rosenthal, J. Kotcher, P. Bergquist, M. Ballew, M. Goldberg, A. Gustafson, and X. Wang. *Climate Change in the American Mind: April 2020*. New Haven, CT: Yale Program on Climate Change Communication, 2020.

Liboiron, Max. *Pollution Is Colonialism*. Durham, NC: Duke University Press, 2021.

Light, Jennifer S. "When Computers Were Women." *Technology and Culture* 40, no. 3 (1999): 455–83.

"Lightning Conductors." *Scientific American* 8, no. 35 (May 14, 1853): 277.

Lipp, J. E., and Robert Salter. "Project Feedback, Summary Report." Santa Monica, CA: RAND Corporation, 1954.

Long Soldier, Layli. "Women and Standing Rock." *Orion Magazine*, 2016. https://orionmagazine.org/article/women-standing-rock/.

Loughran, Trish. *The Republic in Print: Print Culture in the Age of U.S. Nation Building, 1770–1870*. New York: Columbia University Press, 2007.

Loukissas, Yanni A. "A Place for Big Data: Close and Distant Readings of Accessions Data from the Arnold Arboretum." *Big Data and Society* 3, no. 2 (2016): 1–20.

Lucier, Paul. *Scientists and Swindlers: Consulting on Coal and Oil in America, 1820–1890*. Baltimore: Johns Hopkins University Press, 2008.

"Luncheon Address." Presented at the Fifth Annual Technical Exchange Conference of the Air Force Academy, 1969.

Mack, Pamela E. *Viewing the Earth: The Social Construction of the Landsat Satellite System*. Cambridge, MA: MIT Press, 1990.

Maher, Neil M. *Apollo in the Age of Aquarius*. Cambridge, MA: Harvard University Press, 2019.

Mahony, Martin. "For an Empire of 'All Types of Climate': Meteorology as an Imperial Science." *Journal of Historical Geography* 51 (2016): 29–39.

Maini, Anil Kumar, and Varsha Agrawal. *Satellite Technology: Principles and Applications*. 2nd ed. West Sussex: Wiley, 2011.

Malm, Andreas, and Alf Hornborg. "The Geology of Mankind? A Critique of the Anthropocene Narrative." *Anthropocene Review* 1, no. 1 (April 2014): 62–69.

Maphanyane, Joyce Gosata. "Parallel Development of Three Major Space Technology Systems and Human Side of Information Reference Services as an Essential Complementary Method." In *Encyclopedia of Information Science and Technology*, 4th ed., edited by Mehdi Khosrow-Pour, 3484–502. Hershey, PA: IGI Global, 2017.

Markus, Rita M., Nicholas F. Halbeisen, John F. Fuller, James K. Matthews, Joylyn I. Gustin. *Air Weather Service: Our Heritage, 1937–1987: Military Airlift Command Historical Office Special Study*. Scott AFB, IL: Military Airlift Command, US Air Force, 1987.

Marvin, Charles F. *Instructions for Obtaining and Transcribing Records from Recording Instruments, Circular A., Instrument Room, Revised Edition*. Washington, DC: Government Printing Office, 1894.

Marvin, Charles F. *Instructions for Obtaining and Transcribing Records from Recording Instruments, Circular A., Instrument Room, Revised Edition*. Washington, DC: Weather Bureau, 1903.

Marvin, Charles F. *Instructions for Obtaining and Transcribing Records from Recording Instruments, Circular A., Instrument Room, Revised Edition*. Washington, DC: Government Printing Office, 1915.

Marvin, Charles F. "A Weather Bureau Kite." *Monthly Weather Review* 23, no. 11 (1896): 418–20.

McAdie, Alexander G. "Fog Studies on Mount Tamalpais." *Monthly Weather Review* 28, no. 11 (November 1900): 492–93.

McFarland, Stephen L. *A Concise History of the U.S. Air Force*. Washington, DC: US Department of Defense, 1997.

Mergen, Bernard. *Weather Matters: An American Cultural History since 1900*. Lawrence: University Press of Kansas, 2008.

"Meteorology of the Transatlantic Flights." *Bulletin of the American Meteorological Society* 8, no. 8–9 (1927): 113–15.

Middleton, William Edgar Knowles. *Invention of the Meteorological Instruments*. Baltimore: Johns Hopkins University Press, 1969.

Mihesuah, Devon A. *Cultivating the Rosebuds: The Education of Women at the Cherokee Female Seminary, 1851–1909*. Urbana: University of Illinois Press, 1993.

Minnesota Department of Commerce, Energy Environmental Review and Analysis. "Draft Environmental Impact Statement Line 3 Project." May 15, 2017. https://mn.gov/eera/web/project-file?legacyPath=/opt/documents/34079/o.%20DEIS%20Text,%20All%20Chapters.pdf.

"Miscellaneous Inventions." *Scientific American* 50, no. 12 (March 22, 1884): 185–86.

Mitchell, Audra, and Aadita Chaudhury. "Worlding beyond 'the' 'End' of 'the World': White Apocalyptic Visions and BIPOC Futurisms." *International Relations* 34, no. 3 (2020): 309–32.

Mitchell, W. H. "Records by the Kite Corps at Bayonne, N. J." *Monthly Weather Review* 28, no. 12 (December 1900): 539–40.

Moffett, Cleveland. "Scientific Kite-Flying." *McClure's Magazine* 11, no. 4 (March 1896): 379–92.

Monmouth University. "Oral Histories about Super Storm Sandy." *Guggenheim Memorial Library* (blog). Accessed July 5, 2022. https://library.monmouth.edu/oral-histories-about-super-storm-sandy/.

Mott, Frank Luther. *A History of American Magazines, 1741–1930*. Vol. 2. Cambridge, MA: Belknap Press of Harvard University Press, 1958.

"MRCC—Data & Services: 19th Century Weather (FORTS)." Accessed June 26, 2022. https://mrcc.purdue.edu/data_serv/cdmp/cdmp.jsp.

National Advisory Committee on Oceans and Atmosphere. *A Report to: The President and the Congress: National Advisory Committee on Oceans and Atmosphere: Eighth Annual Report, June 30, 1979, Washington, D.C.* Washington, DC: Government Printing Office, 1979.

National Center for Atmospheric Research (US), Advanced Study Program, et al. *Weather Forecasting and Weather Forecasts: Models, Systems, and Users: Notes from a Colloquium, Summer 1976*. Boulder, CO: National Center for Atmospheric Research, 1976.

National Research Council (US) and Space Science Board. *Data Management and Computation*. Vol. 1, *Issues and Recommendations*. Washington, DC: National Academy Press, 1982.

National Resources Defense Council. "Fighting for Safe Drinking Water in Flint." April 13, 2022, https://www.nrdc.org/flint.

Nebeker, Frederik. *Calculating the Weather: Meteorology in the 20th Century*. San Diego: Academic Press, 1995.

New York (State) and Franklin B. Hough. *Proceedings of the Commissioners of Indian Affairs, Appointed by Law for the Extinguishment of Indian Titles in the State of New York. Published from the Original Manuscript in the Library of the Albany Institute*. Albany, NY: J. Munsell, 1891.

Nicollet, J. N. *Essay on Meteorological Observations*. Washington, DC: J. Gideon, Jr., printer, 1839.

O'Brien, Jean. *Dispossession by Degrees: Indian Land and Identity in Natick, Massachusetts, 1650–1790*. Cambridge, MA: Cambridge University Press, 1997.

Olhoff, Anne, and John M. Christensen. *Emissions Gap Report 2019*. UN Environment Programme. November 26, 2019. https://www.unep.org/resources/emissions-gap-report-2019.

Oremus, Will. "The Web's Best Visualizations of Hurricane Sandy." *Slate*, October 29, 2012.

"Ornamental Kites." *Scientific American* 2, no. 3 (October 10, 1846): 24.

"Our Veteran Cooperative Observers." *Monthly Weather Review* 63, no. 11 (November 1935): 313–15.

Papers of Clara Barton. Library of Congress, Washington, DC.

Parkinson, G. R. "Dust Storms over the Great Plains: Their Causes and Forecasting." *Bulletin of the American Meteorological Society* 17, no. 5 (1936): 127–35.

Parkinson, Russell J. "United States Signal Corps Balloons, 1871–1902." *Military Affairs* 24, no. 4 (Winter 1960–61): 189–202.

Parks, Lisa. *Cultures in Orbit: Satellites and the Televisual*. Durham, NC: Duke University Press, 2005.

Pasek, Anne. "Managing Carbon and Data Flows: Fungible Forms of Mediation in the Cloud." *Culture Machine* 18 (2019). https://culturemachine.net/wp-content/uploads/2019/04/PASEK.pdf.

Pawley, Emily. *The Nature of the Future: Agriculture, Science, and Capitalism in the Antebellum North*. Chicago: University of Chicago Press, 2020.

Penny, Virginia. *The Employments of Women: A Cyclopaedia of Women's Work*. Boston: Walker, Wise and Company, 1863.

"Penny Kites." *Scientific American* 49, no. 9 (September 1, 1883): 130.

Perry, Robert. *History of Satellite Reconnaissance*. Vol. 2A, *Samos*. Chantilly, VA: National Reconnaissance Office, 1973.

Perry, Tony C. "In Bondage When Cold Was King: The Frigid Terrain of Slavery in Antebellum Maryland," *Slavery and Abolition* 38, no. 1 (2017): 23–36.

Photo-Miniature, The. New York: Tennant and Ward, 1899.

Pierotti, Raymond, and Daniel Wildcat. "Traditional Ecological Knowledge: The Third Alternative." *Ecological Applications* 10, no. 5 (October 2000): 1333–40.

Pine, Kathleen H., and Max Liboiron. "The Politics of Measurement and Action: 33rd Annual CHI Conference on Human Factors in Computing Systems, CHI 2015." In *Proceedings of the 33rd Annual CHI Conference on Human Factors in Computing Systems*, 3147–56. New York: Association for Computing Machinery, 2015.

Porter, Theodore M. *The Rise of Statistical Thinking: 1820–1900*. Princeton, NJ: Princeton University Press, 2011.

Pryor, Elizabeth B. *Clara Barton: Professional Angel*. Philadelphia: University of Pennsylvania Press, 1987.

RAND Corporation, M. E. Davies, and W. R. Harris. *RAND's Role in the Evolution of Balloon and Satellite Observation Systems and Related U.S. Space Technology*. Santa Monica, CA: RAND Corporation, 1988.

RAND Corporation, Stanley Marshall Greenfield, and William W. Kellogg. "Inquiry into the Feasibility of Weather Reconnaissance from a Satellite Vehicle, R-218," Santa Monica, CA: RAND Corporation, 1951.

Record Group 27. Records of the Weather Bureau. National Archives, College Park, MD.

Record Group 29. *Seventh Census of the United States, 1850*. Microfilm publication M432. Records of the Bureau of the Census. National Archives, Washington, DC.

Record Group 49. Records of the Bureau of Land Management. National Archives, Washington, DC.

Reingold, Nathan. "The Professionalization of Science." In *The Pursuit of Knowledge in the Early American Republic: American Scientific and Learned Societies from Colonial Times to the Civil War*, edited by Alexandra Oleson and Sanborn C. Brown, 33–69. Baltimore: Johns Hopkins University Press, 1976.

Reingold, Nathan. *Science in Nineteenth-Century America: A Documentary History*. Chicago: University of Chicago Press, 1985.

"Report of the Science Advisory Board." Washington, DC: Government Printing Office, 1934.

Report of the Secretary of Agriculture. Washington, DC: Government Printing Office, 1894.

Reports of Meteorological Observations Made at Academies, 1826–1859. Series number A0346-78. New York State Archives, Albany.

Ribes, David, and Steven Jackson. "Data Bite Man." In *"Raw Data" Is an Oxymoron*, edited by Lisa Gitelman, 147–66. Cambridge, MA: MIT Press, 2013.

Rifkin, Mark. *Beyond Settler Time: Temporal Sovereignty and Indigenous Self-Determination*. Durham, NC: Duke University Press, 2017.

Roane, J. T. "Plotting the Black Commons." *Souls* 20, no. 3 (2008): 239–66.

Robinson, W. O. "Composition and Origin of the Dust in the Fall of Brown Snow." *Monthly Weather Review* 64, no. 3 (March 1936): 86.

Rosebud Sioux Tribe and Fort Belknap Indian Community v. United States Department of State; Michael Pompeo, in His Official Capacity; and Thomas J. Shannon, Jr., in His Official Capacity. Accessed June 15, 2022. https://www.narf.org/wordpress/wp-content/uploads/2018/09/keystone-complaint.pdf.

Rosenberg, Daniel. "Data before Fact." In *"Raw Data" Is an Oxymoron*, edited by Lisa Gitelman, 15–40. Cambridge, MA: MIT Press, 2013.

Rosenberg, Daniel. "Early Modern Information Overload." *Journal of the History of Ideas* 64, no. 1 (2003): 1–9.

Rosenzweig, Cynthia, William D. Solecki, Reginald Blake, Malcolm Bowman, Craig Faris, Vivien Gornitz, Radley Horton, Klaus Jacob, Alice LeBlanc, Robin Leichenko, Megan Linkin, David Major, Megan O'Grady, Lesley Patrick, Edna Sussman, Gary Yohe, and Rae Zimmerman. "Developing Coastal Adaptation to Climate Change in the New York City Infrastructure-Shed: Process, Approach, Tools, and Strategies." *Climatic Change* 106, no. 1 (2011): 93–127.

Rotch, Abbott Lawrence. *The Use of Kites to Obtain Meteorological Observations*. Washington, DC: Government Printing Office, 1901.

Ruiz, Rafico. *Slow Disturbance: Infrastructural Mediation on the Settler Colonial Resource Frontier*. Durham, NC: Duke University Press, 2021.

Rusnock, Andrea. "Correspondence Networks and the Royal Society, 1700–1750." *British Journal for the History of Science* 32, no. 2 (1999): 155–69.

Sargent, Jordan. "NASA's Time-Lapse Video of Hurricane Sandy Is Terrifying, Beautiful." *Gawker*. Accessed July 12, 2022. http://gawker.com/5955458/nasas-time-lapse-video-of-hurricane-sandy-is-terrifyingbeautiful.

Schmit, Timothy J., Steven J. Goodman, Daniel T. Lindsey, Robert M. Rabin, Kristopher M. Bedka, Mathew M. Gunshor, John Cintineo, Christopher S.

Velden, A. Scott Bachmeier, Scott Lindstrom, and Christopher C. Schmidt. "Geostationary Operational Environmental Satellite (GOES)-14 Super Rapid Scan Operations to Prepare for GOES-R." *Journal of Applied Remote Sensing* 7 (2013): 1–18.

Schnapf, Abraham, and John H. McElroy. *Monitoring Earth's Ocean, Land, and Atmosphere from Space-Sensors, Systems, and Applications.* Reston, VA: American Institute of Aeronautics and Astronautics, 2000.

Schott, C. A. *Tables and Results of the Precipitation, in Rain, and Snow, in the U.S. and at Some Stations in Adjacent Parts of North America, and in Central and South America.* Washington, DC: Smithsonian Institution, 1881.

Schott, C. A. *Tables of Rain and Snow in U.S.* Washington, DC: Smithsonian Institution, 1872.

Schott, C. A. *Tables of Temperature in U.S.* Washington, DC: Smithsonian Institution, 1876.

Schott, C. A. *Temperature Chart of U.S.* Washington, DC: Smithsonian Institution, 1873.

Second Convention of the Weather Bureau Officials. Washington, DC: Weather Bureau, 1901.

Seitz, Frederick. *The Cosmic Inventor: Reginald Aubrey Fessenden, 1866–1932.* Philadelphia: American Philosophical Society, 1999.

Sellers, Leila. *Commissioner Charles Mason and Clara Barton.* N.p., 1940.

Sharpe, Christina. *In the Wake: On Blackness and Being.* Durham, NC: Duke University Press, 2016.

Signal Service. *Annual Report of the Signal Service.* Washington, DC: Government Printing Office, 1872.

Simmons, Kristen. "Settler Atmospherics." Society for Cultural Anthropology. Accessed June 26, 2022. https://culanth.org/fieldsights/settler-atmospherics.

"Sketch of Professor Coffin." *Popular Science Monthly* 3 (1873): 503–8.

Smith, Heather A., and Karyn Sharp. "Indigenous Climate Knowledges." *WIRE's Climate Change* 3, no. 5 (2012): 467–76.

Smithers, Gregory D. *Science, Sexuality, and Race in the United States and Australia, 1780–1940.* Lincoln: University of Nebraska Press, 2017.

Smithsonian Institution. *First Report of the Smithsonian Institution to the Board of Regents; Giving a Programme of Organization, and an Account of the Operations During the Year.* Washington, DC: Ritchie and Heiss Printers, 1848.

Smithsonian Institution. *Third Annual Report of the Smithsonian Institution to the House of Representatives Showing the Operations, Expenditures, and Conditions of the Institution during the Year 1849.* Washington, DC: Printed by the Printers of the Senate, 1850.

Smithsonian Institution. *Annual Report of the Board of Regents of the Smithsonian Institution: Showing the Operations, Expenditures, and Condition of the Institution.* Washington, DC: Government Printing Office, 1852.

Smithsonian Institution. *Third Annual Report of the Smithsonian Institution to the House of Representatives Showing the Operations, Expenditures, and Conditions of the Institution during the Year 1853*. Washington, DC: Robert Armstrong Printer, 1853.

Smithsonian Institution. *Annual Report of the Board of Regents of the Smithsonian Institution: Showing the Operation, Expenditures, and Conditions of the Institution: For the Year 1854*. Washington, DC: Beverly Tucker, Senate Printer, 1854.

Smithsonian Institution. *Annual Report of the Board of Regents of the Smithsonian Institution: Showing the Operation, Expenditures, and Conditions of the Institution: For the Year 1855*. Washington, DC: Beverly Tucker, Senate Printer, 1855.

Smithsonian Institution. *Tenth Annual Report of the Board of Regents of the Smithsonian Institution: Showing the Operation, Expenditures, and Conditions of the Institution up to January 1, 1856 and the Proceeding of the Board up to March 29, 1856*. Washington, DC: A. O. P. Nicholson, Printer, 1856.

Smithsonian Institution. *Annual Report of the Board of Regents of the Smithsonian Institution: Showing the Operation, Expenditures, and Conditions of the Institution: For the Year 1856*. Washington, DC: A. O. P. Nicholson, Printer, 1857.

Smithsonian Institution. *Annual Report of the Board of Regents of the Smithsonian Institution: Showing the Operation, Expenditures, and Conditions of the Institution: For the Year 1857 and the Proceedings of the Board up to January 28, 1857*. Washington, DC: William A. Harris, Printer, 1858.

Smithsonian Institution. *Annual Report of the Board of Regents of the Smithsonian Institution, Showing the Operations, Expenditures, and Condition of the Institution for the Year 1864*. Washington, DC: Government Printing Office, 1865.

Smithsonian Institution. *Annual Report of the Board of Regents of the Smithsonian Institution, Showing the Operations, Expenditures, and Condition of the Institution for the Year 1869*. Washington, DC: Government Printing Office, 1870.

Smithsonian Institution. *Annual Report of the Board of Regents of the Smithsonian Institution, Showing the Operations, Expenditures, and Condition of the Institution for the Year 1872*. Washington, DC: Government Printing Office, 1873.

Smithsonian Institution. *Annual Report of the Board of Regents of the Smithsonian Institution, Showing the Operations, Expenditures, and Condition of the Institution for the Year 1873*. Washington, DC: Government Printing Office, 1874.

Smithsonian Institution. *Annual Report of the Board of Regents of the Smithsonian Institution, Showing the Operations, Expenditures, and Condition of the Institution for the Year 1878*. Washington, DC: Government Printing Office, 1879.

Smithsonian Institution. *Annual Report of the Board of Regents of the Smithsonian Institution: Showing the Operation, Expenditures, and Conditions of the Institution: For the Year Ending June 30, 1894: Report of the U.S. National Museum*. Washington, DC: Government Printing Office, 1894.

Smithsonian Institution Archives. Record Unit 1. Board of Regents, Minutes, vol. 1. Washington, DC: Smithsonian Institution.

Smithsonian Institution Archives. Record Unit 31. Smithsonian Institution, Office of the Secretary, Correspondence. Washington, DC: Smithsonian Institution.

Smithsonian Institution Archives. Record Unit 60. Meteorological Project, Records. Washington, DC: Smithsonian Institution.

Smithsonian Institution Archives. Record Unit 7001. Joseph Henry Collection. Washington, DC: Smithsonian Institution.

Smithsonian Institution Archives. Record Unit 7060. James Henry Coffin Papers. Washington, DC: Smithsonian Institution.

Smithsonian Institution Bureau of American Ethnology. *Annual Report of the Bureau of American Ethnology to the Secretary of the Smithsonian Institution.* Washington, DC: Government Printing Office, 1895.

Solnit, Rebecca. *The Faraway Nearby.* New York: Penguin Books, 2013.

Sontag, Susan. *On Photography.* New York: Farrar, Straus and Giroux, 1977.

Strakosch, Elizabeth Rose, and Alissa Macoun. "The Vanishing Endpoint of Settler Colonialism." *Arena Journal*, no. 37 (2012): 40–62.

Styres, Sandra, Celia Haig-Brown, and Melissa Blimkie. "Towards a Pedagogy of Land: The Urban Context." *Canadian Journal of Education / Revue Canadienne de l'éducation* 36, no. 2 (2013): 34–67.

Sullivan, Louis H. "The High-Building Question." In *Louis Sullivan: The Public Papers*, edited by Robert Twombly, 76–78. Chicago: University of Chicago Press, 1988.

Swanson, Kara. "Rubbing Elbows and Blowing Smoke: Gender, Class, and Science in the Nineteenth-Century Patent Office." *Isis; An International Review Devoted to the History of Science and Its Cultural Influences* 108, no. 1 (2017): 40–61.

Takarabe, Kae. "The Smithsonian Meteorological Project and Hokkaido, Japan." *History of Meteorology* 9 (2020): 2–23.

TallBear, Kim. "Being in Relation." In *Messy Eating: Conversations on Animals as Food*, edited by Samantha King, R. Scott Carey, Isabel Macquarrie, Victoria Niva Millious, and Elaine M. Power, 54–67. New York: Fordham University Press, 2019.

"Telegraph, The." *Scientific American* 7, no. 42 (July 3, 1852): 330.

"Thoughts for the Working Men." *Scientific American* 1, no. 14 (December 18, 1845): 1.

Tolley, Kim. *The Science Education of American Girls: A Historical Perspective.* New York: Routledge, 2014.

"Tools for Addressing Chapter Conflict." Prepared by Prentis Hemphill with support from the BLM Healing Justice Working Group. Accessed February, 17, 2023. https://accountablecommunities.com/resources.

Trenberth, Kevin E., John T. Fasullo, and Theodore G. Shepherd. "Attribution of Climate Extreme Events." *Nature Climate Change* 5, no. 8 (2015): 725–30.

Tsing, Anna Lowenhaupt. *Friction: An Ethnography of Global Connection.* Princeton, NJ: Princeton University Press, 2004.

Tuck, Eve, and Rubén A. Gaztambide-Fernández. "Curriculum, Replacement, and Settler Futurity." *Journal of Curriculum Theorizing* 29, no. 1 (2013): 72–89.

Tuck, Eve, and Marcia McKenzie. "Relational Validity and the 'Where' of Inquiry: Place and Land in Qualitative Research." *Qualitative Inquiry* 21, no. 7 (2015): 633–68.

Tuck, Eve, and K. Wayne Yang. "Decolonization Is Not a Metaphor." *Decoloniza-tion: Indigeneity, Education and Society* 1, no. 1 (2012): 1–40.

Tufte, Edward R. *The Visual Display of Quantitative Information*. Cheshire, CT: Graphics Press, 1983.

United States and Robley D. Evans. *Wireless Telegraphy. Report of the Inter-departmental Board Appointed by the President to Consider the Entire Question of Wireless Telegraphy in the Service of the National Government*. Washington, DC: Government Printing Office, 1906.

United States and US Civil Service Commission. *Official Register of the United States: Persons in the Civil, Military, and Naval Service, Exclusive of the Postal Ser-vice*. Washington, DC: Government Printing Office, 1903.

United States and Genevieve Walker. *Final Supplemental Environmental Impact Statement for the Keystone XL Project: Applicant for Presidential Permit Trans-Canada Keystone Pipeline, LP*. 2014. Section 2.4.2. http://purl.fdlp.gov/GPO /gpo45808.

United States Congressional Serial Set: Rules and Regulations Governing the Depart-ment of Agriculture in Its Various Branches. Washington, DC: Government Printing Office, 1907.

University of the State of New York. *Annual Report of the Regents of the Univer-sity of the State of New-York: Made to the Legislature, March 1, 1837*. Albany, NY: Croswell, Van Benthuysen and Burt, 1837.

University of the State of New York. *Fifty-Third Annual Report of the Regents of the University of the State of New-York: Made to the Legislature March 2, 1840*. Albany, New York: Thurlow Weed, printer to the State, 1840.

University of the State of New York. *Instructions from the Regents of the University, to the Several Academies Subject to Thier [sic] Visitation*. Albany, NY: Van Ben-thuysen and Burt, 1836.

University of the State of New York. *Instructions from the Regents of the University, to the Several Academies Subject to Thier [sic] Visitation*. Albany, NY: Van Ben-thuysen and Burt, 1845.

US Army Corps of Engineers. "Environmental Assessment, Dakota Access Pipe-line Project." July 2016, 728. https://usace.contentdm.oclc.org/digital /collection/p16021coll7/id/2427/.

US Army Medical Department. "Regulations from the Medical Department." In *Military Laws and Rules and Regulations for the Army of the United States*, 227–28. Washington, DC: W. Cooper, 1814

"U.S. Civil-Service Examination for Assistant Observer, Weather Bureau." *Bulletin of the American Meteorological Society* 1, no. 4 (April 1, 1920): 44.

US Department of Agriculture. *Report of the Chief of the Weather Bureau for 1891*. Washington, DC: Government Printing Office, 1892.

US Department of Agriculture. *Weather Bureau Topics and Personnel*. Washington, DC: Weather Bureau Office, 1915.

US Department of Agriculture. *Weather Bureau Topics and Personnel*. Washington, DC: Weather Bureau Office, 1917.

US Department of Agriculture. *Weather Bureau Topics and Personnel*. Washington, DC: Weather Bureau Office, 1935.

US Department of Agriculture. *Weather Bureau Topics and Personnel*. Washington, DC: Weather Bureau Office, 1936.

US Department of Agriculture. *Weather Bureau Topics and Personnel*. Washington, DC: Weather Bureau Office, 1938.

US Department of Agriculture, Weather Bureau. "Climatological Data: Oklahoma Section." Washington, DC: Government Printing Office, 1930.

US Department of Agriculture, Weather Bureau. *Report of the Chief of the Weather Bureau, 1900*. Washington, DC: Government Printing Office, 1900.

US Department of Agriculture, Weather Bureau. *Report of the Chief of the Weather Bureau, 1901*. Washington, DC: Government Printing Office, 1901.

US Department of Agriculture, Weather Bureau. *Report of the Chief of the Weather Bureau, 1904*. Washington, DC: Government Printing Office, 1904.

US Department of Agriculture, Weather Bureau. *Report of the Chief of the Weather Bureau, 1907*. Washington, DC: Government Printing Office, 1907.

US Department of Agriculture, Weather Bureau. *Report of the Chief of the Weather Bureau, 1908–1909*. Washington, DC: Government Printing Office, 1910.

US Department of Agriculture, Weather Bureau. *Report of the Chief of the Weather Bureau, 1915*. Washington, DC: Government Printing Office, 1915.

US Department of Agriculture, Weather Bureau. *Report of the Chief of the Weather Bureau, 1916*. Washington, DC: Government Printing Office, 1916.

US Department of Agriculture, Weather Bureau. *Special Report of Chief of the Weather Bureau to the Secretary of Agriculture. 1891*. Washington, DC: Weather Bureau, 1891.

US Department of Commerce, Weather Bureau. *Catalogue of Meteorological Satellite Data—TIROS I Television Cloud Photography*. Washington, DC: Weather Bureau, 1961.

US Department of the Interior. *Official Register of the United States*. Washington, DC: Government Printing Office, 1897.

Valenčius, Conevery Bolton. *The Health of the Country: How American Settlers Understood Themselves and Their Land*. New York: Basic Books, 2002.

Viégas, Fernanda, and Martin Wattenberg. *Wind Map*. Site released February 2012 and accessed July 2022. http://hint.fm/wind/.

Voyles, Traci Brynne. *The Settler Sea: California's Salton Sea and the Consequences of Colonialism*. Lincoln: University of Nebraska Press, 2021.

Wark, McKenzie. "The Weird Global Media Event and the Tactical Intellectual [version 3.0]." In *New Media, Old Media: A History and Theory Reader*, edited by Wendy Hui Kyong Chun and Thomas Keenan, 265–76. New York: Routledge, 2006.

"Weather Bureau Men as Educators." *Monthly Weather Review* 34, no. 10 (October 1906): 467.

"Weather Bureau Men as University Students." *Monthly Weather Review* 36, no. 6 (June 1908): 180.

White, Bob. "Diamonds in the Sky: The Contributions of William Abner Eddy to Kiting." *Kite Life Magazine* (October 1, 1999).

Whitnah, Donald R. *A History of the United States Weather Bureau.* Champaign, IL: University of Illinois Press, 1965.

Whyte, Kyle. "Indigenous Climate Change Studies: Indigenizing Futures, Decolonizing the Anthropocene." *English Language Notes* 55, no. 1–2 (2017): 152–63.

Whyte, Kyle Powys. "The Dakota Access Pipeline, Environmental Injustice, and US Settler Colonialism." In *The Nature of Hope: Grassroots Organizing, Environmental Justice, and Political Change,* edited by Jeff Crane and Char Miller, 320–38. Louisville: University Press of Colorado, 2019.

"Wild Man of the Mountains, The." *Scientific American* 1, no. 5 (September 25, 1845): 1.

Williamson, Mark. *Spacecraft Technology: The Early Years.* London: Institute of Engineering and Technology, 2006.

Wilm, Julius. "Poor, White, and Useful: Settlers in Their Petitions to US Congress, 1817–1841." *Settler Colonial Studies* 7, no. 2 (2017): 208–20.

Wilson, Harold S. *McClure's Magazine and the Muckrakers.* Princeton, NJ: Princeton University Press, 1970.

Winston, Jay S. "Satellite Pictures of a Cut-Off Cyclone over the Eastern Pacific." *Monthly Weather Review* 88, no. 9 (September 1960): 295–314.

Wolfe, Patrick. "Settler Colonialism and the Elimination of the Native." *Journal of Genocide Research* 8, no. 4 (December 2006): 387–409.

Worster, Donald. *Dust Bowl: The Southern Plains in the 1930s.* Oxford: Oxford University Press, 2004.

Yusoff, Kathryn. "Anthropogenesis: Origins and Endings in the Anthropocene." *Theory, Culture, and Society* 33, no. 2 (2016): 3–28.

126, 199n32; "neatness" and "untidy" records, 123–24; "plain and unaltered facts," 122–23; recontextualization of data, 147, 155

Cleveland, Parker, 42, 185n42

climate: data, early relation to, 54–55; as settler principle, 56; as word, 28

climate change/crises, 1, 12, 13, 19–20, 174–75; abstraction of data from, 19, 140–41; Bush administration program, 174; data of, 165–69; "emissions gap," 177; lived scale of crisis, 160; positioned as universal and global emergency, 158–59; as product of settler activity, 6, 156, 165–66, 181n28; and settler infrastructure, 156–57, 160, 164; settler narration of, 140, 154, 158, 164–65; as uninterpreted weather reality, 19, 140, 162; visual consumption of, 158–59. *See also* Dust Bowl; Hurricane Sandy

Climate Change Science Program, 174

Cobb Island telegraphy station (Washington, DC), 116

Coffin, James H., 37–40, 49, 51–53, 61, 187nn76–80; *Meteorological Register and Scientific Journal*, 38–40, 40, 51–52; "Report on the Winds of the Northern Hemisphere," 53; "Tables Showing the Prevailing Directions of Winds," 39–40; and women's data labor, 9, 61, 79–83, 85, 192n90

College of New Jersey (Princeton University), 41

colonialism: climate crisis caused by, 166; external and internal, 2, 28, 179n4; militarized, 88, 183n4, 184n14, 201n68; pollution as central to, 56, 181n28; power of, 166, 175. *See also* settler colonialism

Comanche Nation, 126, 201n69, 202n71

Committee on Aeronautical Meteorology, 133–34

Committee on Meteorology (American Association for the Promotion of Science), 54–55

Comsat, 150

Congress, 49, 96; and "Regulations for Conduct of Stations," 119

conquest, 8, 88

consciousness, settler environmental, 17, 21, 89–90, 93, 105–6, 113–14, 135, 156; and display of data, 29–31, 37–38, 51; popular, 22

consumers of data, 9, 17–19, 107, 147, 156–59, 162–66, 177; public settler good, data as, 15, 27–29, 38, 44, 50–51, 55, 103

"Contributions III: Electro-Dynamic Induction" (Henry), 51–52

Conway, Margaret E., 120

Coombs, Crosby, and Eddy Co., 100–101

cooperative observers, 109, 114, 120, 126–32, 200n44; forms used by, 127–29, *128*, *129*

COVID-19 pandemic, 171

Creek (Muscogee) Nation, 60

critical place inquiry, 180n22

Cuban Missile Crisis, 144

currency, data as, 115–16

Dakota Access Pipeline, 151

Dakota and Lakota Nations, 11, 69. *See also* Great Sioux Nation

Daniels, George, 59

Daston, Lorraine, 14, 27, 112, 184n23

data: in 1837 Joint Committee report, 46–48; accessibility of, 51, 55, 174; aestheticization of, 158–63, *161*; big, 9, 14, 174; bodily centered practices, 68–69; climate, early relation to, 54–55; constant/twenty-four-hour streams of, 18, 108, 112, 115, 140, 154, 166; as currency, 115–16; emergence of as concept and practice, 8–10; emergence of in interglacial period, 8; formats, 173; as form of settler colonial world-building, 2–4, 8, 10, 12, 31, 140; as indexical markers, 160–61, 163; as intertemporal and interspatial product, 43; as masses of observations, 21, 42–43;

data (*continued*)

multiregional, 41–43; national concept of, 10, 15, 28, 52, 56, 121, 125, 160, 176–77; observation transformed into, 31, 40–48; overload, at Smithsonian, 9, 78, 80–81; as power, 13–14, 139; as public settler good, 15, 28, 38, 44, 50–51, 55, 103; raw, 14, 118, 149, 152; regional, 15–16, 22, 28, 38, 42, 117, 127, 186n45; as restructuring force, 18, 140; as singular and plural, 21; transmitting, 116–25; trustworthy, 36–37; used by settlers to avoid relationality, responsibility, and accountability, 3, 15, 23, 135, 165–68, 175–77; as visual media, 135–36; weather itself as form of, 12–13. *See also* abstraction/separation of settlers from environment; automated data collection; "clean"/"good" and "dirty"/"bad" data; consumers of data; data labor; data labor, women's; display of data; language of data; numerical data; reduction of data; visual data

data collection: decommissioned satellites as relics of, 172; empire of, 49; eras of, 2–3; as form of power consolidation, 139; hand-based practices, 115, 122–23; two-century settler project of, 2, 9–10, 18, 19, 28, 156–57; vertical thinking about, 135

data cultures, settler, 5–7, 12–16, 21; based on theft of land, knowledge, and life, 48, 51, 56; data as cultural form, 13–14; and satellites, 139–41, 146–47; and upper-air exploration, 89–91, 106–7, 110; and Weather Bureau, 113–14. *See also* "clean"/"good" and "dirty"/"bad" data

data feminism, 13, 58

Data Feminism (D'Ignazio's and Klein), 13

data labor, 9–10, 12, 16–17; caretaking rather than production, 17, 118, 120; categorical flexibility of, 63, 65–66, 68; and white masculinity, 61–62, 75, 83–84, 89

data labor, women's, 16; erasure of, 59, 61,

71; first published account of weather and climate, 84; gaps in feminist and women's studies of, 58; misogyny in shared physical space, 60, 80, 83–85; in semiprivate spaces, 61, 79, 80, 84; women computers, 58–61, 79–86, 192n90, 192n96; women's position in scientific labor system, 84–86. *See also* women, white settler

data regime, 115–16, 132

Davis, Heather, 2, 166

decolonization, 167–69

Defense Meteorological Satellite Program (DMSP), 144

demilitarizing of vision, 145–46, 157

Department of Agriculture, 96, 116, 134

Department of Defense, 139, 142–43, 144, 204n44

De Witt, Moses, 30

De Witt, Simeon, 29–35

Didion, Joan, 10

D'Ignazio, Catherine, 13, 58

disciplining: of data, 16; of observers, 37, 45, 66; "super-vision," 184n23

discreteness, 56

display of data, 14, 29; graphics, 12, 17, 31, 108–9, 121, 157, 177; lexical, 107–8, 183–84n14; numerical form of, 30–31, 37–38; and settler environmental consciousness, 29–31, 37–38, 51; tabular, 30–31, 37

dispossession, 2–6, 30–31, 44–45, 69, 88, 138, 165, 175; and data capture, 51; "progress" as, 156; and science, 69. *See also* Indigenous lands, theft of

domination of environments, 135; weather forts, 4, 88, 193–94n8

Douglas Aircraft Company, 141

Drucker, Johanna, 162

dust: data landscape confused by, 131–32; failure to incorporate into measurement regimes, 115, 131; fields of vision obstructed by, 125–26, 132; not understood as relational, 135; reported by cooperative observers, 127–29, *128*;

travel through precipitation and air systems, 125

Dust Bowl, 13, 18, 125–32, 199n19; as data crisis, 115; as land-use crisis, 125–26

"Dust Storms over the Great Plains: Their Causes and Forecasting," 134

Dutch, 4

Early Modern Royal Society, 27

earthquake, Albany, New York, April 1852, 63, 66–68

Earth Resources Technology Satellite Program, 148. *See also* Landsat

Eastern Shoshone Tribe, 150, 204–5n57

East India Company, 4

Easton, Pennsylvania, women computers, 59–61, 79–84, 192n90, 192n94

ecomedia, 22

Eddy, William Abner (W. A.), 16–17, 89, 95, 97–107, 136, 195–96n45; "Eddy Kite" patented by, 102, *102*; *McClure's Magazine* series, 104; popularity of, 103; self-recording thermograph and meteorograph raised by, 109; War Department, collaboration with, 101–3, 107

education systems, settler colonial, 7, 59–60

Eisenhower, Dwight, 143, 144

electrical experiments, 91–93, 99

"emissions gap," 177

Enlightenment science, 5–6, 10, 27

enumerative strategies, 8, 51, 115, 176

environmental consciousnesses, settler, 14, 37, 113–14, 139

environmental racism, 179n4

environments, as objects to be measured, 5

erasure: and quantification, 6–7; of women's data labor, 59, 61, 71

erroneous data, 112–13, 118–19, 123. *See also* "clean"/"good" and "dirty"/"bad" data

Espy, James, 15, 45, 48–51, 110, 118, 187n72; and Franklin Kite Club, 91; hired by War Department, 48–49; *Writings: First Report on Meteorology to the Surgeon General of the United States*, 48; "The Friends of Science," 48–50; "On the Importance of Hygrometric Observations in Meteorology," 44, 50

"Essay on Meteorological Observation" (Nicollet), 91

Evans, John, 58, 72

experimenters, white settler, 9–10, 16–17, 95; as engine of nation-state, 102–3; in wireless telegraphy, 116. *See also* Eddy, William Abner (W. A.); observers, white settler

extreme weather, 2, 12, 117, 153–54; consumption of, 158–59; settlers distanced from, 158–59, 177

"face of the sky" data recording technique, 39, 57, 68, 174, 176–77, 188n3

factuality, 122–23

Fahrenheit standard, 38

feminism, data, 13, 58

Fergusson, S. P., 99, 100, 109

Fessenden, Reginald Aubrey, 116

Fifth Annual Technical Exchange Conference (Air Force Academy), 137, 143

First Report on Meteorology to the Surgeon General of the United States (Espy), 48

Fixico, Donald, 60

Fleming, James R., 14

folio observation books, 25–26

Forbes, James, 50, 186n66

forecasting, 27, 96, 122, 125, 127; of dust storms, 134; and GOES satellites, 152; Hurricane Sandy, 153–56; "local forecast officials," 114, 119–20; long-range, 133–35; for oil and gas industry, 151; real-time, 161; science of changed by Hurricane Sandy, 154–56; and *Wind Map*, 161; and wireless technology, 117–18. *See also* National Weather Service; Weather Bureau

Foreman, Edward, 63, 78

Fort Gibson, 126, 201n68, 202n71

public patience, 9, 78, 80–81; and Smithsonian Meteorological Project, 51–55, 59, 61, 70–71, 75, 77; women, views on, 76. *See also* Smithsonian Meteorological Project

Henson, Robert, 152

Heymann, Matthias, 112

Hicks, Mar, 58

Holocene, 1

Hopi Tribe, 88, 193–94n8

Hornborg, Alf, 166

Howe, Joshua, 31

Hudson Bay Company, 54

Hui Kyong Chun, Wendy, 158

Hurricane Katrina, 205n76

Hurricane Sandy, 13, 19, 140, 152–60; and Arctic sea ice, 206n105; as crisis of the imagination, 164; as crisis of US settler and imperial states, 166; emergency declared, 158; as entertainment, 155–56; as global media event, 153; science of storm prediction changed by, 154–56

imagination, settler environmental, 3, 10, 15–16, 29, 44–45; crisis of, 164–65; enumerative strategy, 51; and kite technology, 105

Imagined Communities (Anderson), 89

imperial amnesia, 167–68

"improving agriculturists," 4

"In Bondage When Cold Was King: The Frigid Terrain of Slavery in Antebellum Maryland" (Perry), 4

Indian Removal Act of 1830, 44, 201n69

Indigenous environmental knowledges, 6–7, 60, 187n74

Indigenous lands, theft of, 8, 188n2; Anishinaabe, 151–52; Apache and Hopi, 88, 193–94n8; apportionment of occupied lands to whites, 30; Assiniboine, 151–52; Cayuga land, 25, 182–83n1; Creek and Cherokee, 60; Dakota and Lakota, 11, 69; decolonization of, 167–69; Great Sioux Nation, 151–52;

Gros Ventre, 151–52; Haudenosaunee Confederacy, 30, 44; Indian Removal Act of 1830, 44, 201n69; Lenape land, 1, 10, 44–45, 179n2; and meteorological forts, 4, 88, 126, 193–94n8, 201n68; Montana, Wyoming, and Dakota Territory, 88; numericity used for theft of, 26–27, 31; oil pipelines as threat to sovereignty, 151–52; Omaha (UmoNhoN), 57, 188n1; Osage, Kaw, and Comanche Nations, 126, 201n69, 202n71; performance of science tied to, 11–12, 44–45; Ramaytush Ohlone, 9; removal of Indigenous peoples from, 44–45, 201n69; Seminole, 195n28; settler claims over, 2, 167; Shoshone-Bannock and Eastern Shoshone, 150, 204–5n57; Standing Rock Sioux, 151; and surveying, 15, 30–31, 44, 138; treaties, 44–45, 179n2, 183n1, 195n28; "wide extended country" notion, 44–46, 110, 111. *See also* dispossession

Infrared Laboratory (University of Michigan), 204n44

infrastructure, settler, 6, 12, 30, 88, 102–3, 127, 177; climate change and satellites, 155–57, 160, 164; military, 13, 145; oil and gas, 151; vulnerabilities of, 160, 164

injustices, 1–2, 4, 11, 13, 69, 135, 167, 169; alternatives to, 19, 168–69. *See also* dispossession; Indigenous lands, theft of; slavery

In Pursuit of Knowledge: Black Women and Educational Activism in Antebellum America (Baumgartner), 59–60

"Inquiry into the Feasibility of Weather Reconnaissance from a Satellite Vehicle, R-218" (RAND), 141, 143

Institute for Advanced Study, 135

institutions, 3, 10–11, 102–3; collaboration between, 17–20, 135–36, 165, 177; land theft practices, 44–45; and language of data, 15, 28–29. *See also specific institutions*

McClure's Magazine, 104

McKenzie, Marcia, 5, 180n22

memory, computational space as, 157

"merit and discipline," system of, 119–21, 126, 199n32

Meteorological Register and Scientific Journal (Coffin), 38–40, *40*, 51–52

meteorology: categorical flexibility of, 63–66, 68; as holding place for emerging scientific fields, 63–66; military forts built for, 4, 88, 193–94n8

Meteorology Project (Institute for Advanced Study at Princeton), 135

Middleton, William Edgar Knowles, 94

Mihesuah, Devon, 60

militarized colonialism, 88; forts built for meteorology, 4, 88, 193–94n8; photography, use of, 101–4, 107

military, 11; battlefield climatology, 139; disruption missions and bomb-damage assessments, 142; and evolution of TIROS, 141–47; military-meteorological experimentation, 17, 101–3, *102*, 107; and NASA, 138–39, 143–44; reconnaissance, 18, 144–47; weather observers, 46, 183n4, 184n14, 201n68. *See also* TIROS (Television InfraRed Observational Satellite)

mining industries, nineteenth-century, 4

Minnesota Territory, 69

missionaries, 60, 69

Mitchell, Audra, 7

Mitchell, Maria, 71, 72, 80, 190n46

Moderate Resolution Imaging Spectroradiometer (MODIS), 175

Monthly Weather Review, 103–6, *106*, 146–47

Moore, Willis L., 119

Mordecai, Alfred, 46, 47

more-than-human relations, 3, 5, 7, 56, 166, 180n22

MSS (Modular Multispectral Scanner), 148, 150, 204n44

Myer, James, 96

Nadar (balloonist), 97, 101

Nantucket Athenaeum, 71

narration, settler, 14–15, 28, 132, 167–68, 176; about "clean" and "dirty" data, 113–14; of climate crises, 140, 154, 158, 164–65; counter narratives, 168; and data tables, 108; male, 84; and multi-regional data, 41–43; in qualitative observations, 33; and satellite images, 19, 140, 144, 154; textual, 17, 19, 21, 89, 140; and twenty-four-hour data streams, 18

National Aeronautics and Space Administration (NASA), 10, 203n26; demilitarizing of satellites, 145–46; "free and open science," 139, 143, 144, 150–51; Hurricane Sandy video, 19, 152–55; Jet Propulsion Laboratory, 150; military ties, 138–39, 143–44; NASA/Geostat Test Case Project, 149–50; and oil and gas industry, 140, 149–52; and Project 417, 144; and resource extraction, 140, 146. *See also* TIROS (Television InfraRed Observational Satellite)

National Archives, 10

National Centers for Environmental Information, 11

National Digital Forecast Database (NDFD), 159–61, *161*, 163

National Environmental Satellite Center, 173–74

National Oceanic and Atmospheric Association (NOAA), 10, 150, 155, 162, 173; "Scientific Data Stewardship (SDS) Implementation Plan," 174

National Reconnaissance Office (NRO), 144, 145

National Research Council, 173

national science, 3, 8, 18–19, 50–52, 70, 90, 115; national concept of data, 10, 15, 28, 52, 56, 121, 125, 160, 176–77; national weather project, 3, 9–11, 15, 48–56. *See also* science, settler

National Weather Bureau, 13, 17, 18, 20;

and cooperative observers, 126–32; creation of, 95; data turned over to Navy, 116–17; day-to-day forecasting as focus of, 114; inaugural 1890 report, 112; influenced by experimenters, 108; instruction books, 113, 122–24; and kites, 88; masculinizing of, 119–20; "merit and discipline," system of, 119–21, 126, 199n32; meteorological data form (1915), 23; and print culture, 90; and settler environmental consciousnesses, 113–14; and Smithsonian Meteorological Project, 114; station agents, 120–23, 129–31; upper-air stations, 133. *See also* National Weather Service

National Weather Records Center, 173

National Weather Service, 13, 53, 95, 106, 108. *See also* National Weather Bureau

nation-building, 4–5, 29, 43, 56, 62, 89, 108, 146

nation-state, 5, 163–65; bounded by weather data, 43–45, 49, 51, 90, 105, 108, 110; and experimenters, 103–4; and literacy, 89–90; military-meteorological structures of, 13, 138; and satellite technology, 18–19, 138–40, 146, 156–57, 159; shifting concepts of, 62; US settler states as, 6; weather maps used to visualize, 109–10

Nature of the Future: Agriculture, Science, and Capitalism in the Antebellum North, The (Pawley), 4

nature-study movement, 121

"Naval and Military Use of Kite Photography in Time of War, The" (Eddy), 101

Nebeker, Frederik, 199n17

Nebraska Territory, 57, 70

nephoscope, 102

New Deal, 135

"New Instruments in Use at the Weather Bureau," 103

New York State Regents, 15, 29, 39, 187n77; instructions for academies, 31–34; on

occupied lands, 25–26, 31; rules committee, 30–31; "Tables Showing the Prevailing Directions of Winds in the State of New York," 39–40

New York State Surveyor General's Office, 30

New York Times, 88, 92, 94, 101, 103, 146

Nicollet, Joseph, 91

NIMBUS, 18, 140, 147–48, 151, 165

NS-001 Thematic Mapper, 150

numerical data/numericity, 12–15, 29, 44–45, 119n17, 156, 166; advances in, 39–40; display of, 30–31, 37–38; and Dust Bowl, 125–26, 129, 132; enumerative strategies, 8, 51, 115, 176; instrument-based, 107–10; "language of numbers," 8, 108, 187n74; managing, 37; and photographic data, 135–36; pictorial renderings of, 107, 109–10; used for theft of Indigenous lands, 26–27, 31; weather as numerical experience, 12–13; of weather blanks/forms, 20, 32

objectivity, 10, 32, 145, 167

observation: central points for, 52–53, 187n79; managing, 29–48; qualitative, 32–33; relationship of instruments to, 37–38; transformed into data, 31, 40–48; as world-building violence, 31. *See also* sensorial information

observers, white settler: "citizen science," 10–11; climatic data collected by, 127–28; cooperative, 109, 114, 120, 126–32, 128, 129, 200n44; disciplining of, 37, 45, 66; list of, 52–53; "local forecast officials," 114, 119–20; male, 62–68, 64, 65, 66; male, categorical flexibility of, 63, 65–66, 68; military, 46, 183n4, 184n14, 201n68; northeastern United States, 15; self-description of, 26; slaveholders as, 27–28, 62, 69, 73–74, 76–77. *See also* experimenters, white settler

"Of the Influence of Different Winds on the Heights of the Barometer," 48

oil and gas industry, 140, 149–52, 165; hydrocarbon seepage, detection of, 150; and "intraspecies inequalities," 166; "jobbers," 151; pipelines across Indigenous lands, 151–52

Oklahoma weather stations, 126

Omaha (UmoNhoN) Tribe, 57, 188n1

"On the Importance of Hygrometric Observations in Meteorology, and the Means of Making Them with Accuracy" (Espy), 44, 50

Operation Desert Storm (1990–1993), 4

Osage Nation, 126, 201n69, 202n71

otherwise world, 167

Pacific Ocean, 172

Parkinson, G. R., 134

Parks, Lisa, 145–46, 157

Pashley, J. S., 77–78

Patent Office, 55, 78; women computers at, 61–62, 82–85, 192n98, 193n106

patriarchal power, 3, 5, 9, 11, 88–89, 113, 121, 126, 176; brothers and fathers as interface with institutions, 73, 76–78; misogyny in shared physical space, 60, 80, 83–85. *See also* masculinity, white; power

Patrick Draw test site (Wyoming), 150

Pawley, Emily, 4

Pennsylvania land theft, 1, 10, 44–45, 179n2

Perry, Tony, 4

petrocapitalism, 3, 8, 139, 140

Philadelphia, 1

Phinney, S., 25, 28, 29, 30

photography, 109; aerial technologies, 90, 97–98, 107, 157; data visualizations compared to, 160–62; military use of, 101–4, 107; and reality, 161–62; and satellites, 136, 143–46, 150, 153; of urban landscape by kites, 17, 97, 104, 197n73

Photo Miniature, The, 102

pictorial data, 17–18, 107–10, 204n36. *See also* graphical forms; graphs and tables; photography

Pierotti, Raymond, 7

Pine, Kathleen, 13, 14

pollution, as central to colonialism, 56, 181n28

popularity: crises as uninterpreted weather reality, 19, 140, 162; of data, 3, 13, 22–23, 27–28; of Eddy, 103; of tables, 29. *See also* print culture

popular science, 3, 13, 17, 19, 27, 90–91. *See also* print culture

Porter, Theodore, 112

postal service, 59, 60, 127

Potter, Pierpont, 32–33, 32–36

power: of body, 67; challenges to, 6; of colonialism, 166, 175; data as, 13–14, 43, 139; and data culture, 16; languages of, 89; of nation-state, 18–19, 139; and satellite technology, 18–19, 145; of visual knowledge, 145; whiteness as, 11. *See also* patriarchal power

Prattsburgh Academy, 32

Preliminary Design of a World-Circulating Space Ship, 141

Princeton University, 41, 52, 135

print culture, 17, 89–90, 93, 101–2, 105–8; sketch art in publications, 105–6; and "uninitiated" public, 103–4. *See also* language of data; popularity

professionalization, 18, 58–59, 108, 113–14, 198n10; Word War I era, 124–25

programming languages (HTML, Java), 159

"progress," 10, 63, 156

Project 417 (Program 35 and Program 698BH), 144

Project Feed Back, 141

proper nouns/names, 5–6

public data, 29, 46

public settler good, data as, 15, 27–29, 38, 44, 50–51, 55, 103

public sphere, 13, 85

Puerto Rico, 154

satellites *(continued)*
and resource extraction, 18–19, 140, 146–52, 164, 166; and televisual techniques, 141–42. *See also* GOES (Geostationary Operational Environmental Satellites); TIROS (Television InfraRed Observational Satellite)

Schaeffer, George C., Jr., 95

science, settler: amateur and expert networks, 104–5; Enlightenment, 5–6, 10, 27; founded on dispossession, theft, and genocide, 30–31; as within land relations, 56; as masculine, 87–89; performance of tied to land theft, 11–12, 44–45; "progress," 10, 63, 156; and rapid scanning, 154, 156; romanticization of, 92–95; scientific observation, 14, 52; social life of, 15, 29; third-person language used, 26; white women's attempts to claim place in, 61–62, 83–86; "young friends" of, 87–88, 96–98. *See also* national science

Science Advisory Board, 133

Scientific American, 87–88, 101–2, 194n20

"Scientific Data Stewardship (SDS) Implementation Plan" (NOAA), 174

Scientists and Swindlers: Consulting on Coal and Oil in America, 1820–1890 (Lucier), 4–5

Seasat, 150

self-registering instruments, 17–18, 108–11, 114–15, 131; caring for, 121–24; hand calculations in event of failure, 122–23. *See also* automated data collection

Seminole Tribes (Florida and Oklahoma), 95, 195n28

sensorial information, 13, 45–46, 66–68, 118, 140, 164; dust storm reporting, 131; "face of the sky" data recording technique, 39, 57, 68, 174, 176–77, 188n3; tied to capacities of white men, 60

Sergeant, John, 49

"settler atmospherics," 28

settler body: instrumentalization of, 45,

48; male as idealized, 72; male capacities, 60; as site of connection with land, 4–5, 6

settler colonialism: conquest as central to, 8, 88; defined, 180n7; dissolution of settler ways, 168–69; education systems, 7, 59–60; enumerative strategies, 8, 51, 115, 176; imperial amnesia, recognizing, 167–68; and interagency partnerships, 138–39; land, relation to, 5–8; restorative justice needed, 168–69; worldbuilding of, 2–4, 8, 10, 12, 31, 140. *See also* colonialism; data cultures, settler; imagination, settler environmental; Indigenous lands, theft of; infrastructure, settler; narration, settler; science, settler; settlers; violence, settler

settlers: abstraction/separation of from environment, 3, 5–6, 18, 19, 38, 56, 115, 140–41, 146, 158–59, 166, 176, 177; collective forgetting by, 3, 7, 165, 178; and decolonization of Indigenous lands, 167–69; distanced from extreme weather, 158–59, 177; environmental consciousnesses of, 14, 37, 113–14, 139; farmers and teachers, 13; futures imagined by, 14, 31, 113, 138, 167, 198n11; as historic white populations, 2; inability to see themselves, 7, 19, 23, 26, 115, 131–32, 135; infrastructures of white settler perception, 156; literacy of, 89–90, 106; past narrated by, 14, 31; pleasure in aesthetics of environmental horror, 164; reciprocal ties with environmental crises, 115; transformation of being needed, 169

Settler Sea, The (Voyles), 3

"settler time," 156

settler visualization, 17, 27, 50, 90

Sharp, Karyn, 7

Sharpe, Christina, 205n76

Shoshone-Bannock Tribes, 150, 204–5n57

Shugert, S. T., 83, 84

Signal Corps/Service, 95, 96, 127, 133,

theoretical approaches, 114, 125, 133–35

thermometers, 30–32, 35, 38, 39; on kites, 90, 94, 99, 157; self-registering, 111, 129; used by women observers, 57–58, 70, 71, 80

"Third Report of the Joint Committee on Meteorology," 46–48, 47

TIROS (Television InfraRed Observational Satellite), 18, 136, 137–40, 151, 165; aired on television stations, 152; "cloud pictures," emphasis on clarity of, 146–47; command and data acquisition stations, 143; as "global" satellite for public, 146; media reports on, 146–47; military meteorology and evolution of, 141–47; *Monthly Weather Review* report, 146–47; motion picture cameras mounted to, 203–4n32; optical capabilities of downgraded, 143–44. *See also* National Aeronautics and Space Administration (NASA)

Todd, Zoe, 2, 166

traditional ecological knowledge (TEK), 7

Transcontinental Airlines, 134

treaties, land theft central to, 44–45, 179n2, 183n1, 195n28

Treaty of Canandaigua, 183n1

truths, scientific, 27, 42, 47, 56, 147, 156; trustworthy data, 36–37

Tuck, Eve, 5, 168, 179n4, 180n7, 180n22

Tufte, Edward, 163

"undead of information," 158

United States: as "extended country," 44–46, 70, 110, 111; military forts as weather data collection stations, 4, 88, 193–94n8; settler states of as nation-state, 6

universalization, 3, 5, 43, 145, 159

University of Hawai'i, 12

University of Wisconsin, 173

upper-air exploration, 4, 196n47, 197nn73–74; and aviation, 133, 134;

balloons used for, 94–95, 161, 195n33; settler weather data cultures, 89–91, 106–7, 110. *See also* Iraqi Meteorological Department; kites

urban landscape, photography of, 17, 97, 104, 197n73

US Air Corps, 133

US Air Force, 4, 141–42

US Air Force Academy conference, 137, 143

US Air Force Satellite Control Facility (Sunnyvale, California), 145

US Army, 9; Ballistic Missile Agency, 137–38; forts, 4, 88, 126, 193–94n8, 201n68; meteorological data form, 22

US Army Air Force, 141, 143

US Army Electronics Command, 137

US Intelligence Board, 143

US Navy, 53, 54, 139, 183n4; Bureau of Aeronautics, 133; wireless telegraphy owned and operated by, 116–17

Valenčius, Conevery Bolton, 4

variables, and naturalization of separation, 56

Viégas, Fernanda, 159, *161*

violence, settler, 8, 11–12, 28; Anthropocene as signature of settler, 7; collective forgetting of by settlers, 3, 7, 165, 176

vision: demilitarizing of, 145–46, 157; dust as obstruction to, 125–26, 132; as form of weather knowledge, 152; objectivity and sovereignty of, 145; recontextualization of, 147; technics of, 102; technologized, 145–46, 157

visual data, 145; as uninterpreted, 19, 140, 162; visual media, data as, 135–36. *See also* graphical forms; photography

visualization, 203n28; aesthetics of, 158–63, *161*; graphical excellence, rules for, 163; as imprint of reality, 161; real-time, 160–61; settler, 17, 27, 50, 90; ugly environmental data obscured, 163–64 volunteer science, 3, 9–10, 188n1; cooperative network, Weather Bureau, 200n44; "cooperative observers," 109, 114; Dust Bowl weather volunteers, 13; as family activity, 73–74; information sent to Smithsonian, 62–63; list of observers, 52–53; origin stories created by, 15; science culture, 26; by slaveholders, 62, 69, 73–74; Smithsonian Meteorological Project, 9, 51–54; transformed into employee-based systems, 20–21. *See also* experimenters; national science; observers; science, settler

Voyles, Traci Brynne, 3

War Department, 18, 20, 125; and Air Weather Service, 133; automatic processes, experiments in, 108–9; and balloons, 95; Eddy's collaboration with, 97, 101–3, 107; Espy hired by, 48–49; and kites, 88, 97–103, 107–8; Signal Service, 95; and Smithsonian Meteorological Project, 109, 138–39

Wark, McKenzie, 153

waste, 172

water: importance of for life, 1–2; unlocked, used for settler trespassing, 1–4

Wattenberg, Martin, 159, *161*

weather: enslaved people's leverage of, 4, 189n23; as form of data, 12–13; instrumentalized view of, 3, 45, 48; as mixing everything up, 2, 23, 135; as numerical experience, 12–15; as violent uncertainty, 1. *See also* climate change/crisis

weather blanks/forms, 19–20, *20, 21, 22, 23*; dust storms not measurable by, 127–32, *128–31*; form 1001, 111–12, 129–30, 132; lack of uniformity, 31–35; Potter's, 32–33, *32–36*; signatures required, 36–37, 111; and Smithsonian Meteorological Project, 62–64; used by academies, 20,

31–32; women's requests for unheeded, 73–79, 76; yearly journals, 31–34, *33, 34*. *See also* graphs and tables

Weather Bureau. *See* National Weather Bureau

"Weather Bureau Kite, A" (Marvin), 104–6, *106*

Weather Bureau Topics and Personnel, 121

Weather Code Book, 122

Weather Service. *See* National Weather Service

Western Air, 134

western landscapes: agricultural, 4–5; meteorological sites, 88

whiteness: defined, 7; as unnoticed, 11. *See also* masculinity, white; settler colonialism; settlers; specific white institutions and settlers

white supremacy, 11, 60

Whyte, Kyle Powys, 156, 166, 180n7, 181n28

Wildcat, Daniel, 7

Wind Map, 158–63, *161*

Winds of the Northern Hemisphere (Henry and daughter), 80

Wolfe, Patrick, 5, 180n7

women, African American, 59–60

women, white settler: attempts to claim place in science, 61–62, 83–86; brothers and fathers as interface with institutions, 73, 76–78; as cautious participants in meteorological project, 71–72; disadvantaged by "merit and discipline" system, 119–21, 126, 199n32; eliminated from Weather Bureau, 119–20; observers' requests for materials and instruments, 57–58, 73–79, 76; phased out of public work, 83–84; settler logics advanced by, 62, 85–86; as unpaid cooperative observers, 120. *See also* data labor, women's; Patent Office; Smithsonian Meteorological Project